The American Civil Engineer 1852–2002

The History, Traditions, and Development of the American Society of Civil Engineers

FIRST EDITION BY
William H. Wisely, Hon.M.ASCE
Executive Director Emeritus, ASCE

REVISED AND UPDATED BY
Virginia Fairweather
Editor in Chief Emeritus, *Civil Engineering*

American Society of Civil Engineers
1801 Alexander Bell Drive
Reston, Virginia 20191

ABSTRACT: Published in conjunction with the 150th anniversary of the American Society of Civil Engineers (1852–2002), this new edition covers the founding and development of the nation's oldest engineering society. A new, introductory chapter provides an overview of the structure and activities of ASCE today.

Library of Congress Cataloging-in-Publication Data

Wisely, William H. (William Homer), 1906–1982.
The American Civil Engineer 1852–2002 : the history, traditions, and development of the American Society of Civil Engineers / first edition by William H. Wisely ; revised and updated by Virginia Fairweather.
p. cm.
Includes bibliographical references and index.
ISBN 0-7844-0554-9
1. American Society of Civil Engineers. I. Fairweather, Virginia. II. American Society of Civil Engineers. III. Title.
TA1 .W83 2002
624′.06073—dc21 2002026092

Library of Congress Catalog Card No: 2002026092
ISBN 0-7844-0554-9
Manufactured in the United States of America.

Contents

Foreword

Who could have predicted the many events and innovations that have shaped our country and our world over the past 150 years? Certainly not the 12 engineers who met one evening in November 1852 to form what is now the American Society of Civil Engineers. What has remained constant throughout the past century and a half is the role of civil engineers as the stewards of our infrastructure. Everyday they face new challenges to meet the needs of an ever-evolving society.

In the 19th century, civil engineers built the transcontinental rail system linking our still nascent country from coast to coast, and great bridges to serve the growing urban centers. During the 20th century, change continued to accelerate as civil engineers developed safe drinking water systems for all Americans, created the Interstate Highway System that undergirds our country and economy, and turned their attention to solving the environmental problems of an industrialized, urban society. Today, as our country continues to grow and prosper, the warp and woof of the fabric of America is the infrastructure created by civil engineers.

In 2002, ASCE celebrates that day in 1852 when the influence of the Society and its members began to shape our world. To tell the story of the founding, growth and success of the country's first national professional engineering organization, ASCE presents an updated version of *The American Civil Engineer*, originally published in 1974. While the detailed picture painted in the earlier edition remains, we have added a colorful portrait of the past 25 years of challenges and successes of the Society and the civil engineering profession.

We look back at 150 years of glorious achievements of civil engineers and ASCE, and forward at the 21st century full of opportunities. For ASCE, its growing role as the global leader in civil engineering, the changes in knowledge management and delivery, "raising the academic bar" for professional practice, and continuing to serve the civil engineering specialties under one umbrella, present exciting challenges in the decades ahead.

I think the 12 founders would be very proud to see how far the Society has come in developing a network of 125,000 engineers that communicate and work together for the good of civil engineering and society. This book is a tribute to those founding members of the American Society of Civil Engineers and to all the civil engineers who work every day to build a better quality of life for us all.

H. GERARD SCHWARTZ, JR., Ph.D., P.E., F. ASCE
President, American Society of Civil Engineers

American Society of Civil Engineers Foundation

2002: Building The Future

The American Society of Civil Engineers Foundation, Inc. was established in 1994 to assist the American Society of Civil Engineers (ASCE) with resource development and the enhancement of its programs through the philanthropy of its membership and the wider public. The mission of the ASCE Foundation is *to generate resources for the civil engineering profession.*

In 2000, the ASCE Foundation launched the ***2002: Building The Future*** capital campaign to help ASCE and the Civil Engineering Research Foundation (CERF) build strong financial foundations for the future of civil engineering. The monies raised are being used to honor the successes of civil engineering, to increase the public's awareness of the profession, and to help ensure the industry's future value and competitiveness.

The ASCE Foundation gratefully acknowledges the following donors who made gifts of $1,500 or greater to the ***2002: Building The Future*** campaign as of July 10, 2002. Their generous support of ASCE's 150th Anniversary Programs and CERF's Innovation Fund have greatly contributed to the success and legacy of the Society's 150th anniversary, as well as the advancement of the civil engineering profession.

Donors to the 2002: Building The Future Campaign

Titanium Level

$500,000 and Greater

Elizabeth and Stephen Bechtel, Jr. Foundation
Charles Pankow Builders, Ltd.

American Society of Civil Engineers

Platinum Level
$250,000–$499,999

Marie Fletcher Carter, In Memory of Archie N. Carter

AECOM: CTE Engineers, DMJM Aviation, DMJM + HARRIS, Metcalf & Eddy, and Turner Collie & Braden

Gold Level
$100,000–$249,999

Bruce and Joan Coles and Law Gibb Group Inc.
Stephen C. Mitchell

CH2M HILL
Degussa Construction Chemicals, Inc.
MARSH Affinity Group Services a service of Seabury & Smith

Silver Level
$50,000–$99,999

Mr. and Mrs. Robert W. Bein
H.G. Schwartz, Jr.

ARUP

Copper Level
$25,000–$49,999

Mary and Michael Goodkind
Dick and Peggy Karn
William L. and Ellen N. Shannon
Stanley (Mickey) Steinberg

Camp Dresser & McKee Inc.
McGraw-Hill Construction
United Engineering Foundation, Inc.

Bronze Level
$10,000–$24,999

Terrance K. and Betty L. Barry
William J. and Louise J. Carroll
James E. and Kathleen T. Davis
Deepal and Meg Eliatamby
Vi and Bob Esterbrooks
Art and Lorraine Fox
Luther and Lorraine Graef
Irvan F. Mendenhall
Jim and Raeda Poirot
Jan and Chuck Rendall
Ed and Jeanette Robinson
Lawrence H. and Gail W. Roth
Theodore M. Schad
Richard H. Stanley

KCI Technologies, Inc.

Brass Level
$5,000–$9,999

Harl P. Aldrich, Jr.
Louis C. and Susan M. Aurigemma
Harvey and Karen Bernstein
David T. and Monica E. Biggs
Ruth and Walter E. Blessey
Norman L. Buehring
James and Joann Carlsen
DeSales F. and Oliver W. Clemons, Jr.
David L. Collins and PTI Incorporated
Curtis C. Deane
Joseph A. and Eva B. DeFiglia
Melanie Blank and Casey Dinges
Edith R. and John A. Focht, Jr.
Peter Garforth
W.R. "Randy" Gibson
John and Rita Gillespie
Wesley L. and Hideko K. Goecker
Bruce Gossett
Lt. Gen. and Mrs. Henry J. Hatch
Thomas L. and Patricia N. Jackson
Dr. and Mrs. Thomas A. Lenox
Gloria Ma
Charles R. McClaskey
In Memory of Henry L. Mich
David and Janet Mongan
Gerald T. Orlob
Charles A. Parthum
C. Dudley Pratt, Jr.
Dr. and Mrs. Jeffrey S. Russell
E.C. and R.M. Shankland
Paul C. Taylor
Robert and Bernita Thorn
Daniel S. and Linda C. Turner
Carroll and Howard Wahl

Alvord, Burdick & Howson
The Beavers, Inc.
Klotz Associates, Inc.

Florida Section, ASCE
Illinois Section, ASCE

Pewter Level
$1,500–$4,999

Anonymous (3)
John Gordon Ahlers
Jorge Alfonzo-Ravard
Nilda Andrade
Andy and Valentina Andraos
Charles C. Ang
Tung Au
Alice Ann and Dan Barge, Jr.
David Martin Bayer, II
Thomas M. Blalock
George E. Blandford
Carl and Jeanette Blum
Robert F. Borg
Joseph P. Bouquard
Robert L. Brickner
Lauren B. Brown, III
Charles A. Buescher, Jr.
Stephen and Sylvia Burges
Anna Lankford Burwash
John A. Casazza
Jack Edward Cermak
Lloyd T. Cheney
Joanne Y.C. Liou and Cheng Tsuan Chien
Joseph Wing-Kwong Chiu
Katherine and Arthur N.L. Chiu
Richard W. Christie
Ralph E. Compagno
F. Tad Comstock
James H. Conway
Lila and Jose Correa
Wendy and Kenneth Cowan
James J. Craig, Jr.
LeRoy Crandall
John W. Curtis
Nolan H. Daines
Jarir S. and Rihab J. Dajani
Thomas B. Davinroy
Arthur H. Dederman
Rajnikant T. Desai
Umesh Subrao Durgakeri
Charles W. Durham
John E. and Pamela W. Durrant
Dr. and Mrs. Wayne F. Echelberger, Jr.
Donald E. Eckmann
Kenneth L. Edwards
Carol Ann Ellinger
Elbert E. Esmiol
Jon and Kathy Esslinger
Robert N. and Patricia T. Evans
Meggan Farrell
Donald J. Finlayson
John W. Fisher
Jay and Libby Fitzgerald
Eric L. Flicker
Samuel C. Florman
William A. Fortune
Lloyd C. and Marion L. Fowler
Gerard F. Fox
Charles S. Freed
Kenneth G. Fusci
Theodore V. Galambos
Patricia D. Galloway
Carl H. and Ruth Ellen Gaum
William E. Gervasio
Juan A. Gil
David C. Gildersleeve and Thomas H. Gildersleeve, In Memory of Charles L. Gildersleeve
Carl A. Gordon
Moustafa and Nelly Gouda
Wm. Campbell Graeub
John T. Graff
Albert A. and Helene E. Grant
Adriana L. and Edward O. Groff
Leonardo V. Gutierrez, Jr.
Albert E. Haeger
Carl A. Hagelin
Albert H. Halff
Donn E. Hancher
Walter E. Hanson
Yvan Hardy
Heinz Heckeroth
Dave Heldenbrand and Bison Engineering, Inc.
W. Craig Helms
Donald L. and Barbara E. Hiatte
William T. Higgins
Lewis H. Hollmeyer
Russell E. Horn
Richard S. Howe
James C. Howland
Wonsiri Punurai and Cheng-Tzu Thomas Hsu
J.D. Hudak
Donald L. Ifland

Charles A. Irish
Jeanne G. Jacob and Richard G. Frank, In Memory of Clara and Alfred Jacob
James H. Johnson, Jr.
Irving W. Jones
Leland B. Jones
Arthur C. Josephs
Mamoru E. Kanda
Ben Kato
Robin A. Kemper and Family
Paul J. Kendall, In Honor of C.E. Kendall and R.C. Kendall
Conrad and Tywilla Keyes
Jane and Will Kirksey, Honoring Their Parents and Children
Richard W. Kramer
Merrill V. Kreipke
Barbara and Michael Kupferman
Mickey Kupperman
Charles C. Ladd
Bernard L'amarre
Donovan E. Lee
E. Walter LeFevre
Carl and Maria Lehman
Blaine D. Leonard
Thomas M. Leps
Leopold L. Lieberman
Pat and Oscar Lyon
Leroy G. Machala
Yukio Maeda
Donald J. Maihock
Theodore E. Mann
Dennis R. and Catherine M. Martenson
Albert A. Mathews
George and Patsy Matsumura
James and Phyllis McCarty
Fred C. McCormick
James B. McKamey
Aftab Mufti and Zehra Mehdi
Dohn H. Mehlenbacher
John T. Merrifield
William R. Merwarth, Jr.
James Z. Metalios
Henry N. Moore
Khalil-Georges Mrad
Carl W. Muhlenbruch
William G. Murphy, Jr.
James E. Muschell
Dale A. Nelson
Harold L. Nelson
Alan J. Nestlinger
D. Trang Nguyen, In Honor of Her Parents Thuan and Hy-Hy Nguyen
Daniel A. Okun
S. Harry and Margaret Orfanos
R. Stanton Over and Randall S. Over
J. A. "Jay" Padgett, Jr.
Agostino G. Paese
Brian and Ann Pallasch
William H. Palm
C.R. "Chuck" Pennoni
Don Phelps
Robert V. Phillips
Thomas J. Pilch
Charles R. Pittman
Michael A. and Lois G. Ports
Anne Elizabeth Powell
Bobby E. Price
John M. Puckett
Guenther G. Rahm
James A. Reddington
John A. Replogle
Everett V. and Billie K. Richardson
John L. and Jean D. Richardson
Mario Ricozzi
John M. Robertson
Claude G. Robinson
Jerry R. Rogers
Mr. and Mrs. Aldelmo Ruiz-Santiago
Marvin W. Runyan
Gajanan and Sharda Sabnis
Charles G. Salmon
Thorndike Saville, Jr.
Shirley and Tom Sawyer
Mr. and Mrs. Elias Boutros Sayah
Leslie and Howard Schirmer, Jr.
Saul S. Seltzer
Wilbur C. Sensing, Jr.
Francisco J. Serna
Roland and Jane Sharpe
Marisa and Robert Sherard
Stephen P. Sherman
Takayuki Shimazu
Paul and Kathie Skoglund
Delbert M. Smith
Tom and Marcia Smith

Dick and Donna Spencer
Thomas F. Spencer
Thomas F. Stallman
Michael L. Stevens
RADM. and Mrs. Albert H. Stevenson
Bill Stookey
Robert P. Stupp
David and Kristina Swallow
Joseph and Mary Beth Syrnick
John A. Talbott
Daniel Po Kei Tam
Earnest E. Taylor, Jr. and
Taylor Construction Co.
McKinney V. "Mac" Taylor
Nick and Nancy Textor
Alicia Danielle Therrien
Chuck and Karon Tiltrum
Angelo Tomasso, Jr.
John M. Toups
Dennis D. and Jeanie A. Truax
Richard C. Tucker
David W. Turner
Karl E. Voigt
Henry and Alice von Oesen
In Loving Memory of Larry and Judy
Wade
John F. Wagman
Fred P. Wagner, Jr.
Chu-Kia Wang
James C. Webb
Dick and Marge White
John W. White
Robert V. Whitman
Andrew L. Williams, Jr.
Ivan Witkin
John A. Witkowski
Daniel E. Wright
Richard N. Wright
Anna and James Yao
Leonard P. Zick, Jr.

Greenhorne and O'Mara, Inc.
John A. Martin & Associates, Inc.
Leslie E. Robertson Associates, R.L.L.P.

Arizona Society of Civil Engineers,
ASCE
Central Ohio Section, ASCE
Cleveland-Ohio Section, ASCE
Construction Institute, ASCE
Dayton Section, ASCE
Iowa Section, ASCE
Kentucky Section, ASCE
Metropolitan Indianapolis Branch,
ASCE
New Hampshire Section, ASCE
North Carolina Section, ASCE
Tacoma-Olympia Section, ASCE
Tennessee Section, ASCE
Wisconsin Section, ASCE

H. Con. Res. 387 Agreed to June 18, 2002

One Hundred Seventh Congress of the United States of America

AT THE SECOND SESSION

Begun and held at the City of Washington on Wednesday, the twenty-third day of January, two thousand and two

Concurrent Resolution

Whereas,
founded in 1852, the American Society of Civil Engineers is the Nation's oldest national engineering society;

Whereas civil engineers work to constantly improve buildings, water systems, and other civil engineering works through research, demonstration projects, and the technical codes and standards developed by the American Society of Civil Engineers;

Whereas the American Society of Civil Engineers incorporates educational, scientific, and charitable efforts to advance the science of engineering, improve engineering education, maintain the highest standards of excellence in the practice of civil engineering, and ensure the public health, safety, and welfare;

Whereas the American Society of Civil Engineers represents the profession primarily responsible for the design, construction, and maintenance of the Nation's roads, bridges, airports, railroads, public buildings, mass transit systems, resource recovery systems, water systems, waste disposal and treatment facilities, dams, ports and waterways and other public facilities that are the foundation on which the Nation's economy stands and grows; and

Whereas the Nation's civil engineers, through innovation and the highest professional standards in the practice of civil engineering, protect the public health and safety and ensure the high quality of life enjoyed by the Nation's citizens: Now, therefore, be it

Resolved by the House of Representatives (the Senate concurring), That the Congress—

(1) acknowledges the American Society of Civil Engineers for its 150th Anniversary;

(2) commends the many achievements of the Nation's civil engineers; and

(3) encourages the American Society of Civil Engineers to continue its tradition of excellence in service to the profession of civil engineering and to the public.

Attest: *Clerk House of the Representatives.*

Attest: *Secretary of the Senate.*

Preface to the Second Edition

The 150th anniversary of the American Society of Civil Engineers is an important event. What's more, it is an exciting time. The anniversary date comes hard on the heels of the international millennium celebrations, and what better time to sum up the Society's history than after an uninterrupted century of growth and change. Publication of this new edition is just one of many celebratory events marking the anniversary.

ASCE's history from 1852 to 1974 was chronicled by former Executive Director William H. Wisely. His version has been edited to eliminate some redundancies, summarize some lengthy citations, and generally provide a more succinct presentation. A new opening chapter has been added to provide an overview of the Society as it exists today.

Wisely's book is still fascinating history. The early chapters show the parallels between the Society's evolution and the need for engineers in a growing nation. In addition to disseminating knowledge and bolstering the profession in many ways, one early technical accomplishment was taking the initiative on the adoption of standard time in this country, something we take for granted today.

Some parts are surprising, such as the 180-degree turnabout of ASCE policy on professional registration from opposition to encouragement. Discussions about engineering education and the five-year degree go on today. Other things have changed too: the cost of dues in the late nineteenth century and the price of real estate will amuse some readers. Other early ideas will seem quaint, such as the proposal to fine a member who failed to publish a paper or to contribute a scientific document, map, plan, or model to the Society. The fine was to have been $10, equal to the dues in 1880.

ASCE has also helped its members in some very tangible ways. Wisely tells how the Society waived dues for some members during the Great Depression and again for those who served during the two world wars. The Society also helped engineers who were out of work—through efforts by Local Sections to find jobs for their colleagues, and through fund-raising to help those in dire financial straits. He also reports on the Society's efforts with the Engineering Societies Service Bureau, which helped engineers find work and helped establish ASCE's stance regarding the unionization of engineers.

Wisely also discusses ASCE's long tradition of public service, including investigating disasters such as the Ashtabula, Ohio, bridge failure in 1877 and the Tacoma Narrows Bridge failure in 1940. Once again, in the wake of the terrorist attacks of September 11, 2001, ASCE has sponsored teams to investigate the collapse of the World Trade Center towers and the damage sustained by the Pentagon.

ASCE's yearly *Official Register* updates in detail some of the material William Wisely included in his first edition. The ASCE web site (www.asce.org) and the *Official Register* also provide a wealth of easily accessible information.

ASCE has grown and prospered during its 150 years. This book is a tribute to the profession and its professional society, and to their accomplishments. As we enter the new century, ASCE not only continues to provide vast benefits for its members but also to be at the forefront of engineering knowledge and expertise.

VIRGINIA FAIRWEATHER
April 2002

Preface to the First Edition

When the Bicentennial of the United States of America is commemorated in 1976, the American Society of Civil Engineers will have been a witness to 124 years of those two centuries. This historical account, hopefully, will help to relate American civil engineers to the development of his country and his profession throughout this period.

A "Historical Sketch" of ASCE by Secretary Charles Warren Hunt was published as a book in 1897. Mr. Hunt followed this with a sequel in 1917 that was published in *Transactions*, Volume 82, under the title "The Activities of the American Society of Civil Engineers During the Past Twenty-Five Years." In 1947, Professor Edward C. Thoma, M.ASCE, of Purdue University wrote a short review, "Rise and Growth of the American Society of Civil Engineers," which was made available only in a limited edition of mimeographed copies.

The chief aim of this book is to provide an accurate documentary reference to the broad range of ASCE activity since its founding. If the record is more detailed than may be desired by the casual reader, it has been made so in the interest of completeness.

The book is not, however, a chronological recital of events as they occurred. The various programs and service areas of the Society are treated independently, each a story in itself. This is intended to enhance readability and provide better topical access for reference purposes.

Except in the appendix, names of living persons has been almost completely avoided. It is deemed paramount that the thrust of the Society as an entity be unobstructed even by the most impartial effort to identify and judge the roles of the host of individuals who have contributed so much to its work. These evaluations and interpretations are left to other historians.

In an effort to improve readability, specific source references have been limited to a few major items. The primary sources of documentation were the published *Proceedings* and *Transactions*, the annual report of the Society, *Civil Engineering*, and the minutes of the meeting of the Board of Direction. It will not be difficult, therefore, for the reader to pursue a more detailed inquiry in the time frame in which an event is reported.

The hope that the book will be a useful historical reference has been noted. An even more important purpose would be served, however, if it might help both the student and the practitioner to orient themselves to the rich heritage of the civil engineering profession. Present and future leaders of the Society may also benefit from the guidance given here concerning past successes—and failures—in policy and administration. During my work on the book, I have had frequent occasion to regret that the research had not been undertaken before my own service on the staff began in 1955.

The project has been infinitely absorbing and stimulating, and I am deeply obliged to Executive Director Eugene Zwoyer for his continuing interest and encouragement. Many other members of the headquarters staff have responded most generously to my frequent and oft-times impatient requests for data and assistance. Public Information Services Manager Herbert R. Hands and Miss Mary E. Jessup, formerly News Editor of *Civil Engineering*, have been especially helpful in the exhaustive research task that has been necessary. The cooperation of Paul A. Parisi, Director of Publication Services, and Irving Amron, Editor of Information Services, who indexed the book, is noted gratefully.

The book *Engineering and American Society 1850–75* (University of Kentucky Press) was a particularly useful referee, remarkably perceptive in its analysis of the motivation of the early American civil engineer. Its author, Dr. Raymond H. Merritt, Director of Cultural and Technological Studies at the University of Wisconsin–Milwaukee, was kind enough to review this manuscript in detail, and his thoughtful and scholarly criticism was a major contribution.

Highly competent reviews were provided also by Past-President G. Brooks Earnest and Secretary Emeritus William N. Carey, both of whom have been deeply involved with the management of the Society for many years.

Acknowledgment is made to the Department of Civil and Coastal Engineering at the University of Florida—Dr. James H. Schaub, Head, and the office staff collectively—for their tolerance and cheerful cooperation. The staff of the university's Physics and Engineering Library was ever ready to assist, often under trying conditions.

Grateful appreciation is extended to Professor Daniel H. Calhoun, author, and to the MIT Press, publisher, of the book *The American Civil Engineer—His Origins and Conflicts*, for their permission to use a portion of that title for this book.

As she has done so many times in the past 44 years, my dear wife has again indulged me in a professional adventure that has imposed heavily upon my family and home life. This time she has not only graciously forgone much inconvenience and interference with our retirement activities, but she also served as my reviewer and critic. I dedicate this book to her.

WILLIAM H. WISELY
October 1974

CHAPTER 1

The Society Today: A Virtual Transformation

In many ways, the Society that William Wisely chronicled in the first edition of *The American Civil Engineer* no longer exists. Since 1974, ASCE has reinvented itself into a diverse, dynamic organization. This reinvention will continue as the Society seeks to help its members respond to changing world events as well as ongoing technological, political, economic, and educational challenges.

In just one dramatic decade, the 1990s, ASCE moved its headquarters from New York to Reston, Virginia, bought a new building, established the ASCE Foundation, and created the first Institutes. The Society also reached out globally, with many new international liaisons and activities. In addition to these actions, the Society has also launched its own codes and standards program, initiated an ongoing strategic planning process, changed its revenue base to be less dependent on member dues, and become far more proactive in promoting the interests and the accomplishments of the civil engineering profession.

During the past quarter-century, the membership has nearly doubled and grown more diverse. Both staff and budgets to support this growth have increased correspondingly. The electronic revolution has changed not only the way the Society serves it members but continues to dramatically change the profession of civil engineering itself. Change is constant and accelerating, and to keep pace the Society keeps changing.

Now, as the Society celebrates its 150th anniversary and responds to the events of September 11, 2001, the eloquence of ASCE's original 1852 objective—"the advancement of the science and profession of engineering to enhance the welfare of humanity"—is still fresh and relevant.

Writing about the new ASCE and its recent history is a bit like hitting a moving target. As the terrorist attacks of September 11 have shown, events can alter and change even the most recent of histories. What follows is an overview of the recent history and structure of the Society.

New York to Reston

William Wisely records an antipathy to the New York headquarters location that goes back to the nineteenth century. The engineers who saw the need and had the tenacity and vision to found the Society were from New York, and for much of the Society's history the decision-making process was concentrated there. The East Coast was the major U.S. population center at that time and was where most engineers practiced. As the nation's population moved west, the Society's membership outside the New York metropolitan area increased as well.

During the 1970s, the suggestion was first brought to the Board of Direction to move the headquarters out of New York. Costs were a principal concern. High salaries, expensive real estate, and transportation problems were all cited as shortcomings of the New York location. Several different Boards considered the issue of a move and each time voted against the idea.

After considering the work of consultants retained to study relocation options, the Board of Direction voted in April 1993 to move from New York. A search committee was assembled under the direction of David Bayer, and by September 1994 the Society closed on a new headquarters building in Reston, Virginia (Figure 1-1).

The purchaser of the building was the newly established ASCE Foundation, headed and still under the direction of Curtis Deane. The Reston building was purchased for $8 million. Market timing was perfect, for the Society now owned a building that 12 years earlier had cost $16 million to build. In addition, the Society was able to take over an existing mortgage and lease two floors of the six-story building.

The move from the Society's long-time location, the United Engineering Center in New York, was completed during a fifteen-month period from July 1996 through September 1997. A gala grand opening of the headquarters was held in December 1997 to celebrate the new home of the Society, and the building was formally dedicated in April 1998 in conjunction with the Board of Direction's first meeting at the new headquarters.

The ASCE Foundation

As part of the relocation effort, the Board of Direction created the ASCE Foundation in 1994. The Foundation, a federal not-for-profit 501(c)(3) organization, had two initial goals: (1) to purchase, finance, and renovate the new Reston, Virginia, headquarters building for ASCE, and (2) to successfully

Figure 1-1. *ASCE World Headquarters, Reston, Virginia*

conduct a fund-raising campaign to pay for unfinanced needs related to the new building. This capital campaign, "Building for the 21st Century," was highly successful; it raised more than $4.6 million, far exceeding its goal of $3.5 million.

The Foundation Board of Directors includes the President of the Foundation, the current ASCE Executive Director, and the current President, immediate Past-President, and the President-Elect of ASCE. Having successfully completed its initial fund-raising mission, the Foundation embarked upon a strategic planning program, which involved hundreds of civil engineers from around the county. A new mission emerged from this effort: *To generate resources for the civil engineering profession.*

To support this new mission, a number of fund-raising programs were implemented, including a planned giving program, an annual appeal, a sponsorship program for ASCE meetings and conferences, and fund-raising consulting for other nonprofit organizations associated with civil engineering; such as the Civil Engineering Research Foundation (CERF), Chi Epsilon, and the World Engineering Partnership for Sustainable Development. As part of the planned giving program, the Foundation created the Civil Engineering Legacy Society to recognize ASCE members and friends who, by using planned giving vehicles (such as bequests, life insurance or retirement fund designation, and trust or other estate planning gifts) have made a lasting financial contribution to the Foundation.

In October 2000, the Foundation launched its second capital campaign on behalf of ASCE and CERF, "2002: Building the Future," which is the Foundation's commitment to helping ASCE and CERF build strong financial underpinnings for the twenty-first century. The campaign will generate resources to ensure the success of numerous public awareness programs planned in conjunction with ASCE's 150th anniversary and to build an Innovation Fund for CERF to expedite the incorporation of innovations into the practice of the civil engineering, construction, and environmental communities worldwide. The campaign has set a minimum goal of $6 million to be raised over two years, culminating with the ASCE's 150th anniversary in November 2002.

Headquarters for the ASCE Foundation is in downtown Washington, D.C.

The Civil Engineering Research Foundation

ASCE's involvement in research goes back to the nineteenth century. Beginning with its recommendation to President Ulysses S. Grant for testing for iron and steel, through efforts in the 1970s conducted by ASCE Technical Councils on Research, the Society had played a continuing role in the promotion and dissemination of civil engineering research. In the late 1980s, the Board of Direction authorized the creation of the Civil Engineering Research Foundation as a nonprofit organization to focus civil engineering research efforts.

CERF, which began operating in 1989, is charged with bringing together diverse groups within the civil engineering community to facilitate, integrate, and coordinate common solutions to complex research challenges facing the design, construction, and environmental industries. CERF operates innovative technology evaluation centers for highway public works, the environment, and buildings to help industry expedite the transfer of innovation into practice. In 2000, the International Institute for Energy Conservation (IIEC) became a CERF affiliate. IIEC works to accelerate the global adoption of sustainable energy policies, technologies, and practices that foster environmentally and economically sound development. CERF–IIEC has regional offices in Africa, Asia, and Europe as well as project offices in China, India, Latin America and the Philippines.

Through its process of identifying civil engineering research needs and funding sources, CERF develops research plans and coordinates research programs. These research programs may cover any step in the life cycle of a constructed project, including design, construction, operation, maintenance, and planning, and any area where knowledge or technology gaps currently exist and whose solution will advance the state of the art of the industry and the professior.. Technical and management studies may range from bridges and highways, to hazardous waste, to water quality and air quality, to space construction. CERF strives to find better solutions to management and technical problems and to provide the general public with better information.

To carry out its responsibilities and ensure direct participation by the private sector, CERF established the Corporate Advisory Board (CAB). CAB, which is comprised of more than 100 government and industry executives, provides industry direction to CERF activities.

CERF makes an annual report of its activities to the ASCE Board of Direction. CERF is led by Harvey M. Bernstein, current President and Chief

Executive, and it is governed by a 15-member Board of Directors. It is headquartered in Washington, D.C.

The Creation of the Institutes

With the adoption of the 1995 strategic plan, "Working Drawings for the 21st Century," the Board of Direction gave authorization for the organization of Academies or Institutes. Beginning with geotechnical and structural engineering, the Society now has a total of six Institutes (Figure 1-2), with one additional Institute approved for 2002. The creation and development of these Institutes has been perhaps the greatest structural change in the Society since the creation of the Technical Divisions in the 1920s.

At the time of adoption of the 1995 strategic plan, H. Gerard Schwartz, Jr., head of the strategic planning committee said:

> We are in danger of losing major blocks of our membership, or at least their primary allegiance, unless we recognize marketplace realities and make the ASCE the forum for professionals who are involved in par-

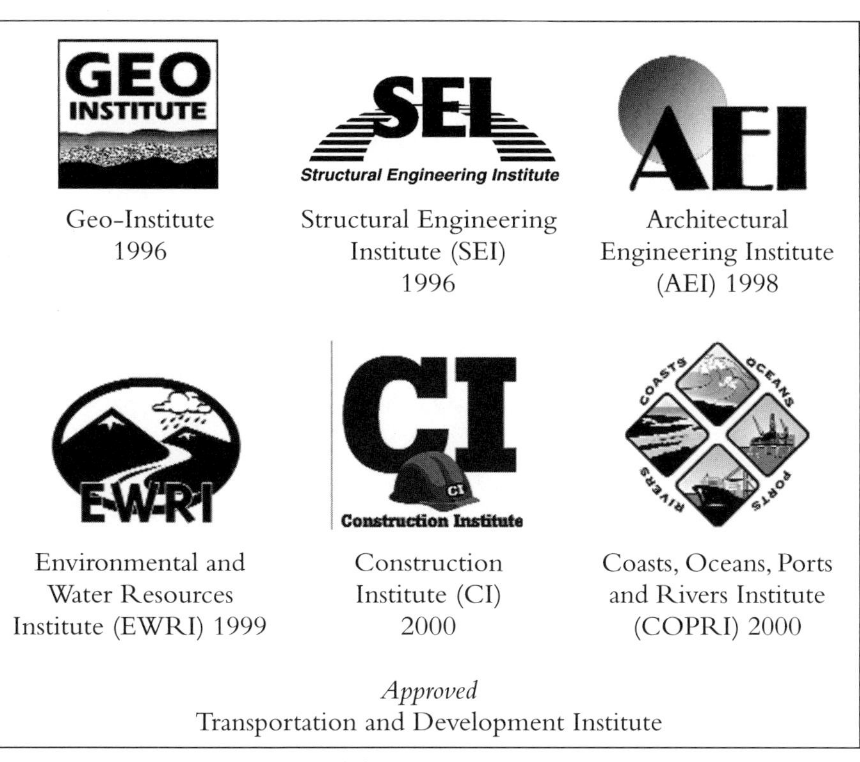

Figure 1-2. *Institutes of the American Society of Civil Engineers*

ticular civil engineering specialties to come together, learn together, and network together. We have lost some specific disciplines already, and the dissolution will continue unless we better meet the needs of members with regard to their areas of specialization and interest.

Membership in the Institutes is not limited to civil engineers; those in "allied professions" may join an Institute, although each has its own membership requirements. ASCE members can become members of one Institute without paying extra fees. Non–civil engineers can become members of Institutes but not of ASCE, although they may be eligible for affiliate membership.

Each Institute has its own bylaws, board of governors, and dues structure and connection to the ASCE Board of Direction is semiautonomous. The Institutes continue to evolve. By 2001, more than half of ASCE members had aligned themselves with an Institute.

Technical Activities Committee

In addition to its growing number of Institutes, ASCE also continues to maintain a wide variety of Technical Divisions and Councils in other areas of civil engineering. The Technical Activities Committee (TAC), an umbrella body encompassing 15 Technical Divisions and Technical Councils, coordinates and supervises all of the Society's technical activities except those specifically assigned to the Institutes. The Codes and Standards Activities Committee also is the responsibility of TAC. A Division focuses on a particular area of technical interest such as air transport or highways, whereas a Council focuses on areas that affect multiple technical areas such as disaster reduction or cold regions engineering.

Each Division or Council (see Table 1-1) is directed by an Executive Committee that monitors the work of administrative and technical committees. For example, in one of the oldest Divisions, the Highway Division, administrative committees include those for awards, publications, and research. The technical committees cover construction and maintenance of highways, environmental quality, geometric design and operations, highway planning and economics, local roads and streets, pavement, and traffic and highway safety. A complete list of the more than 200 committees operating in TAC can be found in the *ASCE Official Register.*

Codes and Standards

Until the mid-1970s, the Society developed codes and standards in conjunction with other organizations, such as the American National Standards Institute (ANSI), the American Concrete Institute (ACI), and the International Organization for Standardization (ISO). Executive Director Eugene Zwoyer proposed that ASCE formulate codes and standards on it own. A staff professional was

Table 1-1. *Divisions and Councils of the Technical Activities Committee*

Division or Council	*Year authorized*
Aerospace Division	1971
Air Transport Division	1945
Technical Council on Cold Regions Engineering	1979
Technical Council on Computing and Information Technology	1973
Council on Disaster Reduction	1997
Energy Division[a]	1922
Engineering Mechanics Division	1950
Technical Council on Forensic Engineering	1985
Geomatics Division[b]	1926
Highway Division	1922
Technical Council on Lifeline Earthquake Engineering	1974
Pipeline Division	1956
Urban Planning and Development Division[c]	1923
Urban Transport Division	1965
Codes and Standards Activities Committee	

[a] Originally authorized as the Power Division.
[b] Originally authorized as the Surveying and Mapping Division.
[c] Originally authorized as the City Planning Division.

hired and a committee was formed consisting of members familiar with ANSI, ACI, and ISO procedures to establish rules for the Society's new program. The process established by ASCE most closely resembles that of ANSI.

In 1978, the Board of Direction approved the establishment of the Codes and Standards Activities Committee. Shortly thereafter, the Society received a U.S. Environmental Protection Agency (EPA) grant to formulate the first ASCE Standard. By 1982, the standards process had grown to seven committees. Today, all activities for the formulation of codes and standards are managed by the Codes and Standards Activities Committee. The work of this committee is divided into three councils: the Lifeline Standards Council, the Geotechnical and Construction Standards Council, and the Water and Environmental Standards Council. By 2001, the Society had published 31 standards, and seven more were nearing the approval stage. Some ASCE Standards are among the best-selling publications of the Society and cover a range of topics—including automated people movers, minimum design loads for buildings and other structures, structural condition assessment of existing buildings, and design and installation of pile foundations.

Information Technology and the Web Site

The first computers came to the Society in the mid-1970s. By 1979, the Society had its first mainframe computer and a permanent information technology (IT) staff of two. Within another decade, the entire headquarters staff had personal computers. The speed of the electronic revolution is underscored by the request of President James W. Poirot to the Board of Direction to approve the purchase of laptop computers for all members of the Executive Committee. The request and subsequent approval by the Board were somewhat controversial and set off a lengthy debate about the advantages and disadvantages of making such a move—a debate that in hindsight seems almost unimaginable.

By 2001, the IT staff had grown to 16 members to support ASCE staff in both the Reston headquarters and the Washington, D.C., offices. The first ASCE web site debuted in October 1996. A complete redesign of the site was completed in 2001. Today, the site (www.asce.org) contains exhaustive information about the Society and its activities and has transformed the way the Society communicates with both its members and the public.

The Board of Direction

The Board of Direction remains the ultimate governing body for ASCE and has had little substantive change since 1974. The Board has 28 members including the officers, the directors representing the Society's districts, and an international member. A District Representative position was created after a 1980 redistricting action in which "representation units" became the basis of seating Board members. Each of the four Zones has five such units. For example, at this writing Zone I has four districts, one of which is twice the size of the other three. Consequently, Zone I's District 1 has two Directors, and the other three have one each. Article V of the ASCE Bylaws details the rotation system that results in a twenty-first Board member who is the District Representative.

In 1999, the Board of Direction transformed its mission. Changes to the Bylaws that year declared that the "Society will have a policy-based Board of Direction, and an oversight-based Executive Committee." The Executive Committee meets more frequently and has the power to make certain decisions on its own. The voting membership of the Executive Committee consists of the President, President-Elect, immediate Past-President, and four Society Vice Presidents, one from each Zone. The Board-appointed Treasurer and Executive Director sit with the committee but do not vote.

In the 1990s, the Executive Committee began holding meetings several times a year to facilitate Board business. Board agendas had become unwieldy so the Executive Committee was empowered to act on items requiring swift official action, such action being confirmed by the Board of Direction at their semiannual meetings. Most committees reporting to the Board of Direction meet twice a year, generally in January and July.

Strategic Planning

Strategic Planning has been used by the Board of Direction to help guide the growth and activities of ASCE since the mid-1980s. The Board appointed the

Task Committee on Strategic Planning in 1983 to prepare a strategic plan for the Society's future. With Albert Grant as the committee's chair (he would become ASCE president in 1988), the first plan was completed and adopted by the Board in 1985.

The first strategic plan contained 91 initiatives grouped into 11 overall goals. Goals were set in education, research, practice, communication, ethics, services, public relations, membership, public involvement, new frontiers, and Society management. In the Executive Summary for the initial plan, the strategic planning process was described as helping the Society to "anticipate the future, rather than react to it, and keep the ASCE in a leadership role in addressing the challenges facing the engineering profession in the years ahead." After submitting its report, the committee was dismissed by the Board because provision had been made for a strategic planning committee to become a permanent standing Board committee.

In 1987, the Board approved an updated version of the initial plan, "From Vision to Action (Edition II)." This revised and updated plan, prepared under Strategic Planning Committee (SPC) chair William H. Taylor, added 44 new initiatives to the original plan. These initiatives were spread among the 11 basic goals set forth in the 1985 plan.

The Board of Direction approved a third strategic plan in 1990. Unlike its predecessors, this plan was described by SPC chair Ed Wilson as "concise, broad in scope, and visionary." Whereas prior plans were long on specifics, this new plan was described as more self-directed. The plan focused on five strategic "thrusts":

1. advance the quality and practice of civil engineering worldwide;
2. expand the civil engineer's involvement in addressing societal issues;
3. improve the development and education of civil engineers;
4. lead infrastructure improvement; and
5. increase member participation and service to members.

The Society's fourth strategic plan was approved by the Board of Direction in October 1995 and given the title "Working Drawings for the 21st Century." Under the direction of SPC chair, and later ASCE President, H. Gerard Schwartz, Jr., this new plan was nearly two years in development and was designed to take the Society into the twenty-first century. In comparison with earlier plans, the publicity and effort devoted to this plan was extensive; articles about the developing plan appeared almost every month in *ASCE News* during 1995.

"Working Drawings for the 21st Century" established both vision and mission statements for ASCE as well as eight specific goals—on infrastructure, global development, policy leadership, recognition, technical excellence, research and development, professional competence, and organization. The boldest initiative of "Working Drawings" was the creation of the current Institutes, with the plan specifically calling for the organization of "academies/institutes" in the field of structural and geotechnical engineering. The plan also provided the direction for creation of ASCE's first *Report Card for America's Infrastructure* as well as the development of the Excellence in Civil Engineering Education project.

The Society's fifth strategic plan, "Building ASCE's Future," was approved by the Board of Direction at the annual conference in the fall of 2000 and prepared under the direction of SPC chair Bruce Coles. The new strategic plan—viewing civil engineers as global leaders in building a better quality of life—set forth ASCE's mission as providing essential value to its members, benefiting society at large, and helping members advance in their careers. Following the precedent of "Working Drawings," this new plan set forth both new vision and mission statements for ASCE. The new vision statement is "Engineers as Global Leaders Building a Better Quality of Life." The mission of the Society is "to provide essential value to our members, our partners, and the public through Developing Leadership, Advancing Technology, Advocating Lifelong Learning, and Promoting the Profession."

Financial and Membership Growth

In 2001, the ASCE annual operating budget exceeded $50 million for the first time, a figure unimaginable just a generation ago. The $50 million includes revenues from membership dues, the Institutes, CERF, and the ASCE Foundation. Though membership dues contributed to the steady financial growth of the operating budget, they now account for less than 24% of the Society's annual operating budget. In a review of the 2000 annual report, publications and advertising contributed the largest share of operating revenues at nearly 30% of the budget; research awards and grants accounted for 15%, continuing education nearly 8%, and meetings and conferences 6%. Current reserves now exceed $25 million, a more than tenfold increase from 1974. A look at the 2000 financial statement (Figure 1-3) shows the diversity of Society operations and financial resources. Through prudent management of its funds and assets, the Society has financially prospered during the course of its 150-year history.

The Board of Direction has the final say on member dues. In 1973, dues were $45 a year for Members and Fellows, with a $10 reduction for Non-Resident Members. An entrance fee of $10 was paid by Associate and Affiliate members. In 2001, the entrance fee was discontinued and the Non-Resident category was abolished. During the past 20 years, there have been 10 increases in dues. In 2002, dues were $175 a year for Members, Affiliates, and Associates. Fellows dues were $210. There are no dues for student members. For recent engineering graduates, a sliding dues scale allows for those one year out of school to pay $50; two years out, $75; three years out, $100; and four years out, $135.

In 1974, ASCE membership stood at 69,671. As of April 2002, total ASCE membership had nearly doubled to 125,886. The membership retention rate

Financials

American Society of Civil Engineers and Affiliates
Consolidated Schedule of Activities
For the years ended September 30, 2001 and 2000

Operating Revenues	FY01	FY00
Membership Dues	13,326,336	11,569,749
Publication Sales	11,718,157	11,026,109
Advertising	3,122,122	3,583,431
Meetings and Exhibits	2,211,009	1,872,239
Speciality Conferences	1,980,993	1,301,860
Continuing Education	4,312,822	3,819,635
Voluntary Contributions	3,015,508	1,988,216
Interest Earned on Investment Excess Cash	324,389	453,063
Investment Income Earned on Marketable Securities	1,600,547	1,710,856
Entrance Fees	38,189	72,609
Royalties	1,759,652	1,121,577
Research and Grants	6,433,124	7,344,537
Label Sales	838,828	893,465
Other Income	1,945,525	1,958,114
Total Operating Revenues	52,619,982	48,706,295
Operating Expenses		
Program Services		
Program Oversite	2,912,939	2,816,474
Membership Services	5,257,479	4,031,013
Publications and Advertising	11,719,087	11,450,980
Meetings and Exhibits	1,614,751	1,545,573
Speciality Conferences	1,963,649	1,755,870
Continuing Education	4,244,001	3,857,138
Educational Activities	1,651,801	1,827,478
Technical Activities	5,983,972	5,910,531
Research and Grants	3,371,162	6,789,624
150th Anniversary	1,207,392	904,466
Government and Public Affairs	2,975,922	2,714,126
Professional Activities	1,146,345	912,345
Total Program Services	44,048,500	44,515,618
Supporting Services		
General and Administrative	5,553,156	2,853,693
Corporate	975,000	933,791
Membership Development	839,852	722,987
Fund Raising	489,459	383,131
Total Supporting Services	7,857,467	4,893,602
Total Operating Expenses	51,905,967	49,409,220
Excess of Operating Revenues Over Operating Expenses	714,015	(702,925)
Nonoperating Revenues and Expenses		
Net Realized and Unrealized Gain (Loss) on Investments	(3,826,454)	825,773
Nonrecurring Expense	(16,971)	(156,837)
Additional Pension Expense	(1,149,805)	0
Total Nonoperating Revenues	4,993,230)	668,936
Increase (Decrease) in Net Assets	(4,279,215)	(33,989)
Net Assets at Beginning of Year	29,881,319	29,915,308
Net Assets at End of Year	25,602,104	29,881,319

American Society of Civil Engineers and Affiliates
Consolidated Balance Sheets
For the years ended September 30, 2001 and 2000

Assets	FY01	FY00
Cash and Cash Equivalents	3,620,530	2,645,718
Investments	23,696,327	28,103,682
Accounts Receivable, Net	3,830,919	4,432,366
Publication Inventory, Net	897,409	959,717
Prepaid Expenses and Other Assets	1,770,555	1,821,600
Pledge Receivable	2,033,777	1,395,082
Fixed Assets, Net	14,742,492	15,136,874
Total Assets	50,592,009	54,495,039
Liabilities and Fund Balances		
Accounts Payable and Accrued Expenses	5,506,125	6,970,113
Unearned Dues and Subscription Revenue	6,420,239	5,088,422
Annual Leave Payable	802,474	710,785
Funds Held for Others	342,269	342,269
Accrued Pension Cost	2,518,798	1,002,131
Bonds and Term Note Payable	9,400,000	10,500,000
Total Liabilities	24,989,905	24,613,720
Net Assets		
Unrestricted	16,774,705	22,554,218
Temporarily Restricted	5,995,114	5,355,158
Permanently Restricted	2,832,285	1,971,943
Total Fund Balances	25,602,104	29,881,319
Total Liabilities and Fund Balances	50,592,009	54,495,039

Revenue Audited Actuals

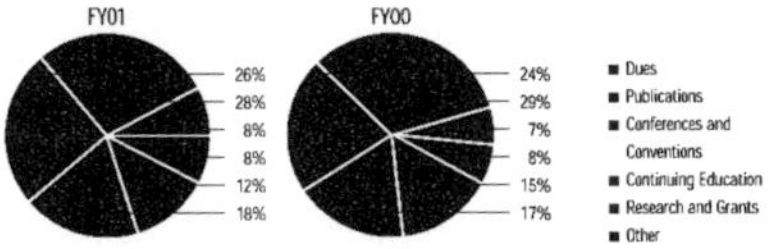

FY01: 26%, 28%, 8%, 8%, 12%, 18%

FY00: 24%, 29%, 7%, 8%, 15%, 17%

■ Dues
■ Publications
■ Conferences and Conventions
■ Continuing Education
■ Research and Grants
■ Other

Expense Audited Actuals

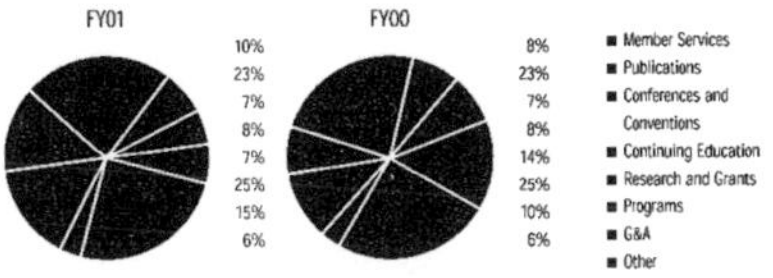

FY01: 10%, 23%, 7%, 8%, 7%, 25%, 15%, 6%

FY00: 8%, 23%, 7%, 8%, 14%, 25%, 10%, 6%

■ Member Services
■ Publications
■ Conferences and Conventions
■ Continuing Education
■ Research and Grants
■ Programs
■ G&A
■ Other

Figure 1-3. *ASCE Financial Statement for 2001*

of 92%—notably high in comparison with other scientific and professional organizations—is a major contributor to the financial stability of the Society. Until 1974, the Society never conducted a membership campaign. In fact, the distribution of membership applications was carefully controlled. However, by turn of the twenty-first century, ASCE had initiated aggressive membership campaigns. The ASCE web site extensively outlined the 2002 campaign; the "Member-Get-A-Member" drive featured membership applications, recruitment tips, and membership benefits. In addition, an extensive list of prizes was available to members who recruited new members.

A Look at the Membership

Membership surveys conducted in 1990 and 2000 as well as surveys conducted for *Civil Engineering* magazine provide profiles of typical ASCE members during the past decade. In 1990, the average respondent was 42 years old, with 16 years of professional experience, and male (94% of respondents). In the 1990 survey, 47% held advanced degrees and the estimated mean salary was $59,776 (compared with $41,521 in a 1983 survey). Of the respondents to the 1990 survey, 22% were structural engineers, followed by 13% for both construction and environmental engineering; 63% of respondents were registered Professional Engineers.

The 2000 member survey asked for slightly different information. The results showed the average ASCE member to be a male in his early 40s with a bachelor's degree and 15 years of ASCE membership. The typical international member was a male in his late 40s with a master's degree and 13 years of ASCE membership.

Over the years, various surveys reported that a principal reason for ASCE membership was keeping up with technical developments through Society publications and conferences. Other reasons cited included the opportunity to develop contacts, the desire to contribute to the profession, prestige, and encouragement as a student to join.

More narrowly defined surveys conducted for the advertising sales staff of *Civil Engineering* magazine provide another look at the membership. In 1999, more than 40% of the members responding were employed in private-practice consulting firms. Another 7.4% worked for architectural/engineering firms and 16.7% worked for federal, state, or municipal governments or regional authorities. Members working for contractors were at 5.5%, engineering educators, 4.8%; and students, 9.2%. Finally, engineers working with producers or equipment and supply firms accounted for 6.7% of members.

Writing about the Society since its inception, William Wisely noted the shift from the proportion of members who worked in railroads in the nineteenth century to private practice and government employment in the twentieth century. This change reflects the advent of the automobile and subsequent public works programs in transportation. Nothing so dramatic has happened to civil engineering employment since. Although Wisely reports cycles in government employment, most members continued to work for private firms.

Each year, a complete list of membership Sections and Branches is published in the *ASCE Official Register*. In 2000, there were 86 Sections and 174 Branches in addition to the District Councils and geographic councils. Each of the four Zones holds "grassroots" meetings of representatives of each Section or Branch each year, a practice begun in the early 1980s at the suggestion of then-President Arthur Fox. In 2001, there were 225 Student Chapters and 38 clubs.

Since 1974, there have been relatively few changes in ASCE membership requirements. In 2000, an ASCE constitutional amendment opened student membership to anyone pursuing a civil engineering degree at a school that has either a student chapter or club; dues for student members were eliminated. The requirements for nomination to the Society Fellow level also changed in 2000. A Section or Branch must make the nomination and the nomination must reflect civic achievement and citizenship in addition to professional achievement.

Diversity

The first edition of *The American Civil Engineer* mentions ASCE's concern with minority representation, describing a challenge by an African-American member to the Society to become more involved with social issues. The Society responded by establishing the Committee on Minority Programs in 1970. Since the mid-1980s, the Society has funded the Summer Institutes, which are designed to attract and prepare high-school-age minority group members to become civil engineers. In 2001, ASCE supported 24 such programs with grants of $5,000 to each program. The Society has also included a voluntary contribution category for diversity and guidance on the dues renewal form.

In 1999, ASCE hired a staff professional dedicated to encouraging diversity within the Society's membership, and in 2000 the Board of Direction created a standing Committee on Diversity and Women in Civil Engineering. The responsibilities of this new committee included promulgating and implementing "programs designed to encourage equitable opportunity for full participation of all people within the civil engineering profession."

According to data voluntarily submitted by those renewing membership, in 2000 there were more than 4,000 African-American members, more than 2,000 Asian members, 2,000 Hispanic members, and 10,000 women members. The voluntary data provide only a glimpse of the diversity of ASCE membership. The Committee on Diversity and Women in Civil Engineering will be encouraging members to provide new information in future membership renewals.

Code of Ethics and Standards of Professional Conduct

The ASCE Code of Ethics, adopted in 1914, was most recently amended in November 1996. The Society is committed to maintaining the highest levels of ethical conduct. All Society members must subscribe to the Code of Ethics, and it is the duty of every Society member to report promptly to the Committee on Professional Conduct (CPC) an observed violation of the Code. The Executive Committee will consider proceedings for the discipline of any Society member upon the recommendation of CPC or upon the written request of 10 or more Society members.

Complete guidelines for separation from membership and disciplinary proceedings are found in the Society's *Official Register.* In addition to enforcing the Code, CPC also sponsors seminars, distributes case studies on ethics for education, and has produced a video on ethical conduct.

The Code of Ethics is also contained within the more comprehensive *Standards of Professional Conduct for Civil Engineers.* The *Standards* were last revised in April 2000. They were developed to provide individuals or small businesses that do not have the resources or a complete set of principles and guidelines to govern the day-to-day aspects of ethics practices within the profession. The five-part document can be found on the ASCE web site (www.asce.org/pdf/ethics_manual.pdf).

Business Development

In 1984, ASCE hired a new staff professional to seek out grants in the civil engineering field. Today, the Business Development Department is a partnership between staff and members to conduct technical projects funded by external sponsors that advance the state of civil engineering practice. Funding sources include federal agencies, private industry, and philanthropic foundations. Staff provides business development, contract administration, and project management support, whereas ASCE members conduct the technical work and are typically compensated for their time and expenses. External funding accelerates the development and delivery of technical products that would otherwise be conducted by volunteer committees. Typical funded projects include pre-standards, guidelines, and manuals of practice; peer reviews; technical reports; field investigations; and conferences, seminars, and workshops. Funding sponsors have included the Federal Emergency Management Administration (FEMA), EPA, Federal Highway Administration, U.S. Department of the Interior–National Park Service, and the Alfred P. Sloan Foundation.

An example of a funded project is the cooperative agreement between FEMA and ASCE for the American Lifeline Alliance public–private partnership. The goal of the Alliance is to reduce the risks to lifelines (utility and transportation systems) from natural hazards.

Engineering Education, ExCEEd, and Continuing Education

In 1998, the Board of Direction considered a proposal that the master's degree be the first professional degree for civil engineers and endorsed Policy Statement 465 stating that the Society "supports the concept of the Master's Degree as the First Professional Degree for the practice of civil engineering at a professional level." The policy proved to be controversial and engendered an outpouring of letters to the editors of ASCE publications. In October 2001, the Board adopted refinements and clarifications of the policy statement and appointed a task committee to work on implementing the revised policy. At that same Board meeting, the title of the policy statement was changed from "First Professional Degree" to "Academic Prerequisites for Licensure and Professional Practice." The new title more accurately reflects the intent of the policy. The Board also acknowledged that implementation of the policy may not occur for 20 years or more.

ExCEEd

Perhaps the most ambitious education undertaking of the Society in recent years is the Excellence in Civil Engineering Education (ExCEEd) project. In 1999, the Society funded summer workshops aimed at raising the standard for civil engineering education, to help faculty improve skills and to create mentoring relationships. ExCEEd was designed, in part, from feedback from the Teaching Teachers to Teach Engineering program of the U.S. Military Academy at West Point, New York. As of this writing, eight weeklong residential programs have been planned through 2002 at West Point and the University of Arkansas in Fayetteville. In the summer of 2000 ExCEEd enrolled almost 200 civil engineering and civil engineering technology faculty in workshops. In addition, the National Science Foundation provided funds for representatives of other engineering societies to observe the program. Each workshop was limited to 24 participants. Participants were screened on the basis of teaching philosophy, administrative support, and a demonstration of how they planned to disseminate what they learned at the workshop.

ASCE has always encouraged practitioners to take part in the education process and has similarly encouraged educators to get field experience. In 2000, a new Practitioner Education Partner Award was created to honor a practicing engineer who participates in education. Candidates demonstrate interest by serving as adjunct faculty, serving on engineering education advisory committees, and otherwise encouraging young students to study civil engineering.

Continuing Education

Although ASCE does not make the pursuit of continuing education a condition of membership, many states now have continuing education as a condition for license renewal. These state requirements have spurred the growth of the Society's continuing education program. In 2000, more than 9,000 individuals participated in an ASCE continuing education course, a number double the enrollment just four years earlier. That same year, ASCE held more than 240 seminars in 45 cities in addition to 40 "in-company" seminars. The continuing education group also offers 75 self-study courses, 14 online courses, and a comprehensive Professional Engineer (PE) Exam Review Course. Continuing education units (CEUs) are awarded for each ASCE continuing education seminar or course. ASCE is an authorized CEU provider and complies with the National Council of Examiners for Engineering and

Surveying continuing education guidelines and International Association for Continuing Education and Training guidelines. A CEU certificate is issued to those successfully completing a seminar or course, and ASCE maintains a permanent record of CEUs earned.

Communications, Awards, and Government Relations

In 1998, ASCE released its *Report Card on America's Infrastructure.* This report evaluated the condition of key components of the nation's infrastructure and awarded an overall grade of D. The report, which was cited twice by President Bill Clinton in speeches that spring, generated a significant amount of publicity for a growing problem for both the nation and ASCE. The report came 10 years after the National Council on Public Works Improvement graded the condition of the U.S. infrastructure, giving it an overall grade of C. Individual grades in the ASCE report were assigned to roads, bridges, mass transit, aviation, schools, drinking water, wastewater, dams, solid waste, and hazardous waste. No category received a grade higher than a C, and the nation's school buildings received an F. ASCE followed up publication of the *Report Card* with a commissioned poll showing the high degree of concern on the part of voters concerning deteriorating infrastructure. The Society also issued a *Voter's Guide to America's Infrastructure.*

In early 2001, the *Report Card on America's Infrastructure* was reissued, adding two new categories, navigable waterways and energy. The overall grade increased only marginally to a D+, compared with the D given in 1998. With the new *Report Card*, ASCE reissued its call for an investment of $1.3 trillion to repair the nation's infrastructure. In issuing the *2001 Report Card*, ASCE President Robert Bein said, "The solutions to these problems involve more than money, but as with most things in life, you get what you pay for. America has been seriously under-investing in infrastructure for decades and this report card reflects that." In 1997, ASCE launched the Congressional Fellows Program, which annually places a civil engineer in the office of a key senator or on a Congressional committee to provide technical assistance on legislative issues related to infrastructure.

The *Report Card* is indicative of the increased efforts on the part of the Society to increase the awareness of civil engineering both among the nation's lawmakers and among the public at large. A 1998 Harris Poll, "American Perspectives on Engineers and Engineering," showed that the nation had a "startling lack of knowledge about engineering involvement in key areas of American endeavor." Further, the poll indicated that a majority (53%) of college graduates were "not very well informed or not at all well informed about engineering and engineers." The Communications and Government Relations section of ASCE has in recent years launched a number of initiatives to increase the public's awareness of civil engineering.

One of the most successful public image programs launched by the Society was the State Public Affairs Grants Programs. In 1996, the Board of Direction approved a program that funded government relations and public relations efforts at the state and local levels. Since 1996, nearly $600,000 in grants has been awarded. Almost every state has participated in the program, which on average has awarded 75 grants each year. Sections and Branches submit proposals for grants, which are screened by the Managing Director of the Communications and Government Relations Department as well as two members of the Board. Grants have been awarded to fund engineering career days, legislative days at state capitals, reports cards on state and local infrastructure, National Engineering Week activities, and a wide range of other activities that attract media attention or promote legislation.

In 2000, ASCE cosponsored *Building Big*, a Public Broadcasting Service (PBS) documentary television series on civil engineering and construction. The five-part series, which was developed and narrated by noted author and illustrator David Macaulay, used archival footage and expert interviews to tell the story of the world's civil engineering marvels. The programs aired on more than 300 PBS stations and garnered more than 18 million viewers. As part of a *Building Big* educational outreach program, ASCE members visited schools in 10 communities nationwide, educating more than 20,000 students about the engineering principles explored in the series.

Sponsorship of *Building Big* was the first project of the Society's 150th anniversary celebration. This 15-month-long celebration, which will culminate in the annual conference and exposition in November 2002 in Washington, D.C., is coordinated by the ASCE Foundation. Other activities of the celebration include sponsorship of a float in the annual Tournament of Roses Parade, cosponsorship of the West Point Bicentennial Bridge Engineering Design Contest, and the sponsorship of several museum and traveling exhibits.

With the dawn of the new millennium, ASCE canvassed its members in late 1999 to determine the 10 civil engineering achievements that had the greatest impact on life in the twentieth century. Rather than individual projects, they chose to recognize broad areas of achievement (Table 1-2). Following this "Millennium Challenge," a prestigious panel of civil engineers selected one international or national project to represent each of the 10 achievements. These "Monuments of the Millennium" (also listed in Table 1-2) demonstrate a combination of technical engineering achievement, courage, and inspiration, and they have had a dramatic influence on the development of the communities in which they are located.

Another successful effort in government relations has been the ASCE Key Contact program. Initiated by the Government Relations office in the mid-1980s, the program encourages and facilitates relationships between individual members and government officials. Former ASCE President Albert Grant, himself a government employee, took the lead in encouraging civil engineers to recognize their professional responsibility to be involved in government. The Key Contact program is a result of that initiative.

Table 1-2. *The 10 Greatest Civil Engineering Achievements and Monuments of the Millennium*

Achievements	*Monuments of the Millennium*
Airport design and development	Kansai International Airport
Dams	Hoover Dam
Interstate highway system	The Interstate Highway System
Long-span bridges	The Golden Gate Bridge
Rail transportation	The Eurotunnel Rail System (linking Britain and the rest of Europe)
Sanitary landfills/solid waste disposal	—
Skyscrapers	The Empire State Building
Wastewater treatment	Chicago Wastewater System
Water supply and distribution	The California Water Project
Water transportation	The Panama Canal

Note: As of this writing, a Monument of the Millennium has not been selected for sanitary landfill/solid waste disposal.

When the Key Contact program began, the goal was 5,000 ASCE Key Contact members. At this writing, there are more than 15,000 Key Contact members. Members in the program receive a weekly email newsletter, "This Week in Washington," or a monthly fax newsletter, "The Month in Washington." Both provide information about federal and state legislative and regulatory issues affecting civil engineers. Key Contact members also receive periodic "Key Alerts" about issues that need the immediate action of civil engineers to educate legislators about the engineering perspective. Key Contact program members have contacted Congress about issues such as infrastructure financing, dam rehabilitation, environmental policy, natural hazards impact mitigation, and education investment.

The Communications and Government Relations office also monitors the efforts of relevant regulatory agencies such as the Office of Federal Procurement Policy and EPA. Through this office, ASCE also offers advice to federal agencies in appointing qualified engineers to federal positions.

Publications

The Society's publication goals changed radically in the late 1970s when then–Executive Director Eugene Zwoyer carried out recommendations of what was known as the "Yoder Report," named for Charles Yoder, an ASCE Past-President. The report called for decreasing the Society's reliance on dues revenue and finding other sources of income. At that time, dues accounted for 83% of ASCE's revenues. A director of publications and marketing director were hired and charged with making the publications operation self-sustaining. Since the 1980s, publications have been a major source of income for the Society. Despite this success, the Board of Direction in 1996–97 actively considered the sale of this asset. However, the process of preparing a sales prospectus clearly showed the Board the value of publications to the Society, and the idea of a sale was ultimately rejected. In fiscal 2000, publications sales and advertising accounted for nearly 30% of the Society's annual revenues. By contrast, in that same year, dues accounted for approximately 24% of revenue.

The activities of the Publications Division are wide ranging. The Society's flagship publication, *Civil Engineering*, was first issued in 1930 and remains a major source of advertising revenue. The magazine, which is published monthly, informs both the membership and the civil engineering community in general of advances and issues in the profession.

In 1976, *ASCE News* debuted as a monthly newspaper focusing on the Society's activities and member achievements. *ASCE News* replaced many individual Technical Division newsletters that had been published prior to 1976. Subscriptions to *Civil Engineering* and *ASCE News* are a benefit of membership.

A major focus and source of growth of the Publications Division are the 29 journals published under the direction and guidance of the Professional Activities Committees and the Technical Divisions, Councils, and Institutes. Online access is available to all subscribers. In addition, all 29 journals are collectively available on a CD-ROM that is published quarterly.

In 2000, the Publications Division initiated the *Personal Journal*, which (for an annual fee) allows subscribers to select and download 25 articles from any of the Society's journals online. Scheduled for publication in 2003 is an innovative new electronic journal, *Civil Engineering Letters*. This journal will make brief, cutting-edge communications in all disciplines of civil engineering rapidly available to engineers around the world.

Visitors to the ASCE web site can search the Civil Engineering Database, an electronic index of all journals, papers, feature articles, and books published by the Society since 1972. Users can download abstracts at no charge or purchase individual articles from a growing archive of the Society's journals (1995–2002).

ASCE also publishes an extensive array of other publications for the Society and Institutes, including standards; manuals and reports on engineering practice; committee reports; and proceedings in print and CD-ROM formats. ASCE Press, a publishing imprint of the Society, is dedicated to the publication of scholarly authored and edited works on technical subjects, civil engineering management, regulatory compliance, and the history and heritage of civil engineering. The popular *Bridges* calendar, first published by the Communications Department in 1980, has become an annual Society tradition. The Publications Division also publishes annual editions of the *Official Register* and *Transactions of the American Society of Civil Engineers.*

Society Awards

The honors program of the Society has as its basic objective the advancement of the engineering profession by emphasizing exceptionally meritorious achievement. Traditionally, such accomplishments have been made in the form of research papers, although many awards are based on other considerations. The awards are made by the Board of Direction, in the name of the entire Society, on the recommendation of Society agencies designated in each particular case. As of 2001, the Society bestowed 79 different awards. This is more than double the number awarded in 1974. The awards ceremony is normally held at the Annual Meeting. To accommodate the presentation of so many awards, the ceremony was converted to a multimedia presentation during the presidency of Arthur Fox. Award winners are listed each year in the *Official Register.*

Two additional awards presented by the Society each year are for the Outstanding Civil Engineering Achievement (OCEA) and the Outstanding Projects and Leaders (OPAL). OCEA, which was established in 1960, recognizes an exemplary civil engineering project in the United States. The award honors the project that best illustrates superior civil engineering skills and represents a significant contribution to civil engineering progress and society. By honoring an overall project rather than an individual, the award recognizes the contributions of many engineers. Past award winners include the Cape Hatteras Light Station Relocation Project, the Denver International Airport, and the Ted Williams Tunnel in Boston. A complete list of OCEA award winners can be found in the *Official Register.*

OPAL was established in 1999. This award recognizes and honors outstanding civil engineering leaders whose lifetime accomplishments and achievements have made significant differences in design, construction, public works, education, or management. OPALs, considered the "Academy Award" of the civil engineering profession, are awarded each year to recipients in each of the five categories. A complete list of OPAL award winners can be found in the *Official Register.*

The Society Staff

When ASCE was founded in 1852, a sole unpaid Secretary officiated. When William Wisely wrote the first edition of *The American Civil Engineer* in 1974, there were 109 employees. When Executive Director Eugene Zwoyer took office in 1972, he initiated organizational charts, salary guidelines, and systematic performance reviews. In 2001, the Society had more than 300 full- and part-time staff, including those working in Reston and the Washington, D.C., offices for ASCE, CERF, and the ASCE Foundation.

In 1974, the Human Resources Department (HR) was a one-person operation. In the past 25 years, the number of employment laws has grown exponentially. Today, HR has six employees who oversee compliance with a wide range of employment regulations and requirements. The HR staff is also responsible for recruiting, benefits administration, employment policy development, employee training, and managing the performance appraisal process. HR also produces the employee newsletter, *The Society Page.*

As with any dynamic organization, organization charts and assignments are fluid and changing as events demand. A complete description of the organizational structure is given each year in the *Official Register*. Major operational units of the Society include the ASCE Foundation; Communications and Government Relations; Corporation and General Counsel; CERF; the Institutes; Educational and Technical Activities; Publications; Continuing Education; Conferences; Membership; Professional Activities, Geographical Services, and Diversity; International Activities; and Knowledge Management.

Since 1974, the staff has been led by several Executive Secretaries and Directors. Eugene Zwoyer served as Executive Secretary from 1972 until 1982. He was succeeded by Edward Pfrang, who served from 1983 to 1994. James E. Davis has served as Executive Director from 1994 to the present.

The Society's International Reach

The international reach of the Society has accelerated dramatically in the past 25 years, with dedicated staff to support international activities first appointed in 1992. In 1985, the "Agreement of Cooperation" was established to facilitate communication and joint activities among engineering organizations worldwide. The Agreement creates a mutual relationship among professional engineering organizations to stimulate the exchange of technical, scientific, and professional information. The goal is to learn from and exchange with international colleagues and unite the engineering profession worldwide. As of late 2001, ASCE had entered into 69 Agreements of Cooperation in 55 countries. A complete list of these Agreements can be found in the *Official Register.*

In 2000, ASCE signed the *Edinburgh Accord* with the Institution of Civil Engineers (ICE) of the United Kingdom to facilitate the two organizations' need to meet challenges posed by the rapid changes in the engineering marketplace worldwide. Through this accord, the world's two leading civil engineering societies pledged to work together to improve the quality of practice and the professional tolls for civil engineers throughout the world. This dynamic and far-reaching partnership—which offers the benefits of reciprocal memberships—envisions using the Internet to create a global knowledge network and establish a gold standard of professional qualification that would serve as an international passport of engineering competence and a vehicle for international professional mobility. As a virtual partnership of organizations working in harmony with other engineering institutions, ASCE and ICE will provide professional services to support the needs of more than 500,000 practicing engineers throughout the world.

In 2000, the Board of Direction passed a global resolution expanding on the strategic plan's vision statement regarding engineers as global leaders. Noting that ASCE began its collaboration with international counterparts over a century ago, the resolution expressed the Society's intention to grow its international membership, hold conferences throughout the world, and use the Internet to further the advancement of civil engineering. Also in 2000, the Board created an International Affiliate membership for those who belong to the professional engineering society of their own country.

9/11: The Society Responds to a National Tragedy

The terrorist attacks of September 11, 2001, in New York and Washington, D.C., will likely influence the work and focus of ASCE and its members for years to come. Like the rest of the nation, ASCE members and staff watched in horror as the events of that unforgettable day unfolded. The collapse of the twin towers of the World Trade Center and the severe damage to the Pentagon presented immense challenges both to the nation and to the Society.

Civil engineers were among the first on the scene at both disasters, initially to provide assistance in rescue and recovery efforts and later to determine what lessons could be learned from the destruction at both sites. Within days of the attacks, ASCE, using its Disaster Response Procedure, authorized the formation of two teams to study the collapse of the twin towers and the damage to the Pentagon.

The World Trade Center Data Collection Team was lead by W. Gene Corley, Ph.D., P.E., a senior vice president with Construction Technologies Laboratory in Skokie, Illinois. Dr. Corley, an expert in building collapse investigations, was also the leader of the forensic team that investigated the Murrah Federal Office Building in Oklahoma City following its bombing in 1995. The entire World Trade Center Data Collection Team consisted of 22 members offering a comprehensive range of structural engineering experience. The Society and the Structural Engineering Institute helped to coordinate the participation of a coalition of engineering societies, including the American Institute of Steel Construction, American Concrete Institute, Council on Tall Buildings and Urban Habitats, International Code Council, National Fire Protection Association, Society of Fire Protection Engineers, Structural Engineers Association of New York, and the National Council of Structural Engineering Associations. In May 2002, FEMA published *World Trade Center Building Performance Study,* prepared by the ASCE team.

Paul F. Mlaker, PhD, P.E., led the Pentagon data collection team. Dr. Mlaker, a leading expert in the design of structures to resist explosive effects, was at the time of his appointment the technical director of the Research and Development Center of the U.S. Army Corps of Engineers. He was also a member of the ASCE forensic team, which assessed the performance of the

Murrah Building in Oklahoma City. Mlaker's Pentagon team consisted of five leading structural and fire protection engineers. Publication of the team's report is expected in Summer 2002.

In addition to the formation of data collection teams, the Board of Direction also approved the establishment of a critical infrastructure response initiative (CIRI). With this initiative, the Society will play a leading role in addressing infrastructure vulnerability and in developing strategies and guidelines for mitigating the effects of natural and human-made disasters on the critical elements of the nation's infrastructure. As outlined by the Board, the initiative will seek to:

- assess infrastructure vulnerability;
- prioritize infrastructure renovation based on the results of vulnerability assessments;
- determine research and development directions that will help to protect critical elements of the infrastructure;
- develop retrofit designs to mitigate disaster damage;
- formulate new design procedures to mitigate disaster damage; and
- improve disaster preparedness and response procedures.

In addition to analyzing the performance of buildings, CIRI will examine the vulnerability of other infrastructure systems—among them those for water supply, transportation, waste management, energy, and telecommunications.

On the Horizon

What lies ahead for the ASCE? In his inaugural address in October 2001, H. Gerard Schwartz, Jr., President of the Society in its 150th anniversary year, looked ahead. The events of September 11, 2001, he said, "have shaken our confidence in our economy, our invincibility, and perhaps our democracy. As civil engineers, what role do we play in this war on terrorism? How do we use our professional skills to, as our canon of ethics states, hold paramount the safety, health, and welfare of the public?"

The new President pointed out that when ASCE celebrated its 100th anniversary, "we were just recovering from World War II, a great Iron Curtain had fallen across Europe. The Cold War and the Korean War had begun. Communism versus capitalism. Totalitarianism versus capitalism. Was it any less certain or frightening than today?" "Yet," said Schwartz, "in the face of such monumental challenges, look at what we accomplished over the next 50 years. With steadfast determination, we built a better country and a better world, for all its flaws."

In addition to the Society's response to September 11, Schwartz also addressed other objectives that were no less important to the long-term health and vitality of ASCE and the profession of civil engineering. Leadership is a key element of ASCE's strategic plan, he said, and "the objective is to restore civil engineers to leadership roles in our communities and nation. We need to create, or re-create, civil engineering as a profession of leaders."

Schwartz noted recent Board actions to revise and clarify the policy on the educational requirements for being licensed as a civil engineer. "We will raise the bar, the academic bar for the professional practice of civil engineering in the twenty-first century," he said. "To do otherwise would be to countenance a steadily eroding role for the civil engineers of the future."

Schwartz also said that the Society was well along in discussions to incorporate a widely recognized specialty certification academy into ASCE. Efforts were also under way to establish specialty certification academies within at least three of ASCE's Institutes. Schwartz also said that ASCE has to be at the forefront of the new information technologies that are revolutionizing the practice of engineering. However, he warned, "Let us be careful to use information technology to enhance wisdom, not substitute for it."

In closing his address, the Society's new president said, "As we launch our one hundred fiftieth anniversary, let us look back with pride at what our profession has achieved, but look forward also to dreams not yet built, to projects not yet conceived. For we are the builders of America and the world."

CHAPTER 2

American Engineering as a Profession

The American Society of Civil Engineers, the first professional engineering society in the United States, began life in 1852 after almost 30 years of sporadic effort. The fledgling group was barely established when the Civil War eclipsed its activities. Shortly after the war was over, ASCE emerged in 1867 as the modest symbol of the new profession of engineering.

Across the Atlantic in England, engineering became an official profession when the British Institution of Civil Engineers was created in 1818, the culmination of efforts going back at least to 1771. National engineering organizations were formed in Holland (1843), Belgium (1847), Germany (1847), and France (1848).

Philadelphia was the starting point for engineering organizations in the United States. In 1789, the Society for Promoting the Improvements of Roads and Inland Navigation was the earliest known organization focused on public works engineering. Civil Engineer William Strickland visited England on behalf of this society in 1825, and his 1826 report described canals, locks, bridges, roads, tunnels, tramways, breakwaters, harbors, and other public works.

Considering the magnitude of public works needed in this country during the nineteenth century, and the thin ranks of those capable of providing them, it is not surprising that organizing an engineering society seemed less important than getting on with the work. The first half of the nineteenth century was a crucial period in U.S. history. Citizens had migrated as far west as the Mississippi by 1820. Development was largely rural, and 90% of the national population of 10 million lived in settlements of fewer than 2,500 residents. Western farms were devoted to food crops, Southern agriculture was dedicated to cotton, and manufacturing made a modest beginning in New England and some larger Western cities.

To integrate these activities into national trade and commerce, communication and transport were sorely needed. Engineers played important roles in many bold and imaginative enterprises that provided these services and engineers began to take a nonmilitary role during this period, mainly in transportation. Early civil engineers became leaders in planning, building, and managing bold public works projects that shaped the nation.

The first law providing federal funds for nonmilitary public works was enacted in 1787. The statute established the Northwest Territory, decreeing that 2% of revenues from land sales be allocated to building the National Road or "Cumberland Pike" from Washington to the Ohio River. Begun in 1806, the first 130 miles to Wheeling, Virginia (now West Virginia), were opened in 1820, and the road eventually ran to Columbus and Saint Louis. The 9,000 miles of rock and gravel surfaced roads in 1820 grew to 88,000 miles during the next four decades.

Equally significant was the short-lived Canal Era. The South Hadley and Middlesex canals in Massachusetts (1793 and 1804) were followed by the greatly successful Erie Canal in New York (1825), the Union Canal in Pennsylvania (1829), the Morris Canal in New Jersey (1831), and the Chesapeake and Ohio Canal (1851). By midcentury, the steam engine was in general use for water and land transportation. About 3,000 miles of canals were operating in 1840, by which time the waterway was rapidly yielding its leading role in transportation to the railroad.

The South Carolina Railroad was the first in the United States, followed by the Baltimore and Ohio; the Pennsylvania; the Pittsburgh, Fort Wayne, and Chicago; the Rock Island; and the Erie railroads. The Erie was a training ground for seven engineers who later became ASCE presidents between 1868 and 1895: James P. Kirkwood, William J. McAlpine, Horatio Allen, Julius W. Adams, William E. Worthen, Octave Chanute, and George S. Morison. By 1852, the early American railroad system covered about 9,000 miles. That number increased tenfold in the next 30 years.

The railroad system had a great impact on bridge construction; until the early nineteenth century, bridges were largely built with timber and stone masonry. The middle years of the century brought a renaissance in bridge building, both in the transition from wood to cast and wrought iron and in the advancement from the early empirical trussed arches to the proprietary trusses of Howe, Pratt, Warren, Whipple, Bollman, Fink, and others. The systematic analysis of stress in the truss members and early efforts toward rational design soon followed. The first metal truss bridge (cast and wrought iron) was designed and built on the Reading Railroad in 1845. By midcentury, wrought iron was in general use as a structural material, and the arrival of the Bessemer converter in 1856 introduced the Steel Age.

In another area of civil engineering, there were 83 municipal water supplies in the United States in 1850, thanks largely to pioneering efforts in Boston, New York City, and Bethlehem, Pennsylvania. These suppliers provided untreated water from wells, springs, and surface sources. The earliest American water treatment was the use of slow sand filters in Saint Louis in 1866.

Although some sewer lines in Boston were operating by 1800, the earliest public system of sewers was built in Chicago in 1855 under the direction of City Engineer E.S. Chesbrough. By 1860, public sewer systems were serving about a million people in 10 of the largest cities. Sewage was discharged without treatment to the nearest watercourse.

Civil Engineering Defined

The term "civil engineer" was first used formally in Great Britain. John Smeaton, builder of early roads, structures, and canals, signed himself under that title in presenting expert testimony in the British courts around 1782. In the United States, the Continental Congress legislated the appointment of engineer officers in the army, and most of these positions were filled by Europeans. This "Corps of Engineers" was disbanded at the end of the Revolutionary War, then established again in 1794. In 1802, Congress created the U.S. Military Academy at West Point, New York, as an arm of the Corps of Engineers, and it remained so for more than 60 years.

In 1821, Congress enacted legislation directing the Army Corps of Engineers to make surveys of major roads and canals, and it prescribed that this work be performed by office and field parties under the direction of a supervisory board, all to be jointly constituted of "engineer officers" and "civil engineers." This is certainly one of the earlier distinctions in the United States between the military engineer and the civilian or "civil" engineer.

The U.S. Military Academy was an important source of trained engineers. Of the 572 graduates from 1802 to 1829, 49 became chief or resident engineers on railroad or canal projects by 1840. Most formally educated engineers of this period were West Point graduates. Others supplemented their general academic training by scientific study and field experience. Many engineers in the mid-1800s, however, had little or no formal education. They acquired their technical knowledge through self-study and apprenticeship, often as axmen or rodmen in surveying parties. The roads, canals, and railroads on which they worked were their universities.

West Point's curriculum was greatly strengthened during the tenure of Sylvanus Thayer as Superintendent from 1817 to 1833. The first civil engineering course outside of West Point was offered in 1821 by the American Literary, Scientific, and Military Academy, later called Lewis College, and then, in 1834, renamed Norwich University. Rensselaer Polytechnic Institute conferred the nation's first civil engineering degree in 1835. By midcentury, engineering courses were being offered by Union College (1845), Harvard College (1846), and Yale College (1846). In the next 20 years, about 70 institutions of higher learning initiated engineering programs.

The national census of 1850 counted only 512 civil engineers, of whom two-thirds lived in Massachusetts, New York, Ohio, Pennsylvania, Connecticut, and Wisconsin. Although few in number, civil engineers were an elite group. They were supremely confident of their capabilities, and they jealously guarded their independence of professional decision and action. As champions of the public interest, they were outspoken in their criticism of questionable political and industrial management. Their performance earned these civil engineers prestige, public respect, and financial remuneration to a degree not matched by any other profession at the time.

The United States in the mid–nineteenth century was a brash young nation, rich in natural resources, a diverse blend of rugged frontiersmen and immigrant artisans in its population of 24 million. The government offered unprecedented opportunities for individual innovation and private enterprise, and the Industrial Revolution was well under way. The Singer sewing machine, Goodyear rubber, Colt firearms, and the McCormick reaper were among the foremost contributors to the London Exposition in 1851.

It was against this backdrop that the leaders among U.S. professional builders began to think and talk about an organization that would provide a facility for their communication and concerted action in matters of technical and professional concern. The time could not have been more propitious!

Early Disappointments

When the Franklin Institute was established in Philadelphia in 1824, it was the professional home of many engineers, and for some years its journal published much of the American and foreign engineering literature. Dr. J. Elfreth Watkins wrote in 1891 about the earliest U.S. effort in 1836 to create a "National Society of Civil Engineers." This attempt, by engineers with the Charleston and Cincinnati Railroad, "did not meet with the encouragement expected, and the society was short-lived."

Late in 1838, "a highly respectable meeting of members of the profession in Augusta, Georgia," resulted in a call for a convention of civil engineers in the United States to be held in Barnum's Hotel in Baltimore on February 11, 1839. The Baltimore convention was held, "forty gentlemen of the profession being present, from the States of Massachusetts, New York, New Jersey, Pennsylvania, Illinois, Maryland, Virginia, Missouri, North Carolina, Georgia and Louisiana." Benjamin H. Latrobe Jr. was elected President of the convention, and John Frederick Houston was appointed Secretary.

Forty practitioners attended the convention, at least 10% of all U.S. civil engineers in 1939. The assembly adopted the following resolution:

> Resolved, that the convention now proceed to the election of a committee of seventeen, to prepare and adopt a constitution and form a Society of Civil Engineers of the United States.

In accord with a decision that "the different portions of the Union may be represented" in the committee, those elected to serve were Benjamin Wright, New York; William S. Campbell, Florida; Claude Crozet, Virginia; W.M.C. Fairfax, Virginia; C.B. Fisk, Maryland; Edward F. Gay, Pennsylvania; Walter Gwynn, North Carolina; J.B. Jervis, New York; Jonathan Knight, Maryland; Benjamin H. Latrobe Jr., Maryland; W.G. McNeill, South Carolina; Edward Miller, Pennsylvania; Moncure Robinson, Virginia; J. Edgar Thomson, Georgia; Isaac Trimble, Maryland; Sylvester Welsh, Kentucky; and G.W. Whistler, Connecticut.

The Franklin Institute of Philadelphia offered space for headquarters, meetings, and a library; a secretarial service; space for publication of papers in the Institute's journal; and even editorial staff.

The Baltimore convention also authorized appointment of "a committee of five to draft an address to the Civil Engineers of the United States, and to superintend the publication of such portions of the proceedings of this convention as they may deem expedient." Appointees to this committee were C.B. Fisk, Isaac Trimble, J.B. Jervis, G.W. Whistler (later replaced by Edward Miller), and S.W. Roberts of Pennsylvania.

The Committee of Five promptly convened on March 20, 1839, to draft the statement for circulation to all civil engineers in the Union, which stated in part:

> Public works are now so extended … and the mass of experimental knowledge … so peculiarly applicable to our circumstances, that it is even more valuable to the American Engineer than what he can learn in Europe. … In this country … [it is critical] to obtain the greatest amount of useful effect at the smallest cost. … The student, or the more advanced Engineer … seeks in vain for any satisfactory written or printed description, and is unable to obtain anything more than vague … and incorrect information. …
>
> A society in this country must differ somewhat in its plan of operations from the British Institution, which can readily give utterance to its opinions elicited after frequent and full discussion, since a large portion of its members … have their residences within the limits of London. Here, however, owing to the vast extent of territory over which are scattered the members of our profession, the usefulness of the Society must (for the present at least) depend more upon the facts and experience of its members, made known in written communications, than upon their opinions orally expressed in public discussions.
>
> The difficulty of meeting at any one point, caused by the time and expense required in traveling from distant portions of so extensive a country as the United States, is a serious obstacle. … The standing of the profession in our country is, fortunately, such that its importance need not be dilated upon. …

> The Committee will close this address by a quotation from the inaugural address of the distinguished Thomas Telford, the first President of the London Institution, which appears to them peculiarly appropriate:
>
> "In foreign countries similar establishments are instituted by government, and their members and Proceedings are under its control, but here a different course being adopted, it becomes incumbent on each individual member to feel that the very existence and prosperity of the institution depends, in no small degree, on his personal conduct and exertions; and the merely mentioning the circumstance will, I am convinced, be sufficient to command the best efforts of the present and future members, always keeping in mind that talents and respectability are preferable to numbers, and that from too easy and promiscuous admissions, unavoidable and not unfrequent incurable inconveniences, perplex most societies."

When the Committee of Seventeen met on its appointed date of April 10, 1839, only four of its members attended: Messrs. Wright, Campbell, Fisk, and Miller. They drafted a constitution for "The American Society of Civil Engineers," to be "instituted for the collection and diffusion of professional knowledge, the advancement of mechanical philosophy, and the elevation of the character and standing of the Civil Engineers of the United States."

Although the Member grade was to be limited to persons who are or have been engaged in the practice of a civil engineer, an Associate grade would have accommodated architects and "eminent Machinists, and others, whose pursuits constitute branches of Engineering, but who are not engineers by profession." Another provision imposed a fine of $10 against any member who failed "to produce to the Society at least one unpublished communication in each year, or present a scientific book, map, plan or model, not already in the possession of the Society."

The limited participation of members in the April 1839 meeting of the Committee of Seventeen was an omen. Only four other committee members (Robinson, Welsh, Latrobe, and Jervis) endorsed the draft constitution. It was never submitted to another convention for adoption, and the Franklin Institute's generous offer of support went begging.

The Secretary of the Committee of Seventeen, Edward Miller, gave his explanation for the collapse of the movement in a communication published in the *American Railroad Journal* in February 1840. In his opinion, the Committee of Seventeen was too large for its purpose, in addition to "the facts that most of those appointed were ignorant of their appointment, several absolutely indifferent or hostile to the formation of any institution; and that many were unknown to each other and so scattered as to render a meeting difficult." He noted further that "under these conditions there can hardly be a necessity for pointing out the local views, partialities and jealousies which influenced in some measure the result."

Miller also concluded that jealousy and discontent must result from the necessity of vesting the management in a few who must reside near the point at which the Society's hall is located. Four independent regional societies, each with its own headquarters, was his answer to this provincialism. His recommendations were never followed, but in the years to come there were several manifestations of regional and intraprofessional bias. Miller's disappointment at the collapse of the 1839 organizational movement comes through clearly in the following letter he addressed to the members of the Committee of Seventeen:

Philadelphia, July 15, 1839

Sir:

I have the honour to inform you that the form of Constitution proposed for the Society of Civil Engineers by that portion of the Committee of Seventeen which met in Philadelphia on the 10th of April, agreeably to their appointment, is rejected.

The votes are as follows:

Approving—Benjamin Wright; Wm. S. Campbell; Charles B. Fisk; and M. Robinson. Edw. F. Gay also approves of the Constitution, but declines becoming a member of the Society under any circumstances.

Disapproving—W.M.C. Fairfax; Walter Gwynn; John B. Jervis; Jonathan Knight; B.H. Latrobe, Jr., W.G. McNeill; J. Edgar Thomson. C. Crozet expresses no opinion on the subject, but declines his appointment as one of the Committee of Seventeen.

Isaac Trimble; Sylvester Welsh; and G.W. Whistler, have made no reply to my circular letter, and I am not acquainted with their views.

From the tenor of the letters received from the different members of the Committee, I have been convinced that a *National Society* on a broad and useful basis, can not be formed by gentlemen holding such discordant opinions, unless they will take the pains to meet together, and give the subject a fair discussion. I am also of opinion that a subject of such importance should not be decided by a meagre majority, and therefore although still believing the Constitution to be a good one) I add my vote to the negatives, which *will* make the names against the proposed measure eight, Mr. Crozet having withdrawn from the Committee. The proposed Constitution is of course *rejected,* even though all the remaining gentlemen to be heard from should vote in its favor.

I have hitherto cheerfully attended to the duties which the Baltimore Convention, and subsequently the Committee imposed upon me, but must confess that I now see no prospect of a beneficial result to the profession. So I have no leisure for useless correspondence, I respectfully decline acting longer as the organ of the Committee, and will hand over the papers and correspondence, in my hands, to any one whom they may designate.

Very respectfully, Edw. Miller

The discussions of "the highly respectable meeting of members of the profession in Augusta, Georgia," came to a disappointing end.

The Watkins paper also refers to an abortive move in Albany, New York, in 1841 to establish "The American Institute of Civil Engineers." In 1848, local activity resulted in the formation of the New York Institute of Civil Engineers and the Boston Society of Civil Engineers. The New York group published a *Transactions* volume but ceased activity in 1850 for lack of support.

The Boston Society of Civil Engineers (BSCE), set up in 1848, was the first permanent engineering organization in the United States. The organization included only two states and the concentration of engineers in Massachusetts and Connecticut eliminated some of the problems a national organization faced. Other likely reasons for its success were the rigid admission requirements that limited membership to mature, less mobile and more prestigious individuals; the immediate establishment of a headquarters for meetings and a library; and the group's inclusion of social as well as professional activities. BSCE served its regional membership with distinction for 154 years. In 1974, it affiliated with ASCE, but it still operates under the name of the Boston Society of Civil Engineers. James Laurie, one of the founders of BSCE, later became a Founder and the first President of ASCE (Figure 2-1).

Figure 2-1. *James Laurie, the First President of the Society*

The American Society of Civil Engineers and Architects, 1852

In October 1852, this notice was sent to practitioners of civil engineering in and near New York City:

> New York City, October 23rd, 1852
>
> Dear Sir:
>
> A meeting will be held in the office of the Croton Aqueduct Department, Rotunda Park, on Friday, November 5th, at 9 o'clock P. M. for the purpose of making arrangements for the organization, in the city of New York, of a Society of Civil Engineers and Architects.
>
> Should the object of the meeting obtain your approval, you are respectfully invited to attend.
>
> Wm. H. Morell, Wm. H. Sidell, J.W. Adams, A.W. Craven, James Laurie, James P. Kirkwood, and others.

Twelve respondents gathered at the appointed time in the office of Alfred W. Craven, chief engineer of the Croton Aqueduct Department. They were Julius W. Adams, J.W. Ayres, Alfred W. Craven, Thomas A. Emmet, Edward Gardiner, Robert B. Gorsuch, James Laurie, W. H. Morell and W.H. Sidell of New York; G.S. Greene of Albany; S.S. Post of Oswego, and W.H. Talcott of New Jersey.

With Craven presiding, the group resolved to incorporate an American Society of Civil Engineers and Architects, and designated Laurie, Adams, and Sidell to draft a Constitution. The committee swiftly produced a draft Constitution that was discussed, amended, and adopted at once. This constitution remained in force until 1868.

The first elected officers were: James Laurie, President; Edward Gardiner and Charles W. Copeland, Vice Presidents; J.W. Adams, A.W. Craven, James P. Kirkwood, William H. Morell and William H. Sidell, Directors; and Robert B. Gorsuch, Secretary and Treasurer.

The 12 individuals who gathered in Rotunda Park on November 5, 1852, merited the appellation of "Founders." This was the contention of an unidentified member of the original 12 in a letter responding to an article published in the *Engineering News* of March 8, 1890. The letter observed: "No one could be considered a 'founder' of the Society, who, in view of the published call to engineers, had virtually declined, by neglecting either to wish us 'God-speed,' or to personally participate in the honors attending the leading of what was so generally regarded as a forlorn hope." The editor of *Engineering News* admitted the validity of the premise, and apologized for past inaccuracies in referring to "founders" of the Society.

The death announcement for Colonel Julius Walker Adams in the December 16, 1899, issue of the *Engineering Record* alluded to his role in ASCE's formation. Referring to Colonel Adams's service as editor of Appletons' *Mechanic's Magazine and Engineer's Journal*, the notice states: "It was while he held this office that the famous Wacamahaga, a half-social, half-technical club, was formed by Mr. Adams, Henry R. Worthington, Charles W. Copeland, James O. Morse, James How, C.M. Guild and others, who later organized what is now the American Society of Civil Engineers." The name "WA-CA-MA-HA-GA" likely came from the first initials of some members' names. During 1852, the club's proceedings, published by the *Mechanic's Magazine and Engineer's Journal*, show that Colonel Adams was a Founder of the Society; Copeland and Morse, Charter Members; and How and Worthington, Members. Alfred W. Craven, another Founder, was also a WA-CA-MA-HA-GA member.

The 1852 organizational meeting at Croton called for the adoption of an "address" to be distributed to all engineers and architects in the United States considered eligible for membership, to invite their affiliation. This statement follows:

> It has been for some time under advisement to form in the city of New York a Society of Civil Engineers, embracing also the kindred professions, with a view to their mutual improvement and the public good. Accordingly a meeting was called on the evening of the 5th of November of such professional men as were accessible and were supposed to be favorably inclined to such an association. The objects of the contemplated Society were laid before this meeting, as also the means by which it was proposed to accomplish the end in view. A Constitution was drawn up, discussed in detail, and finally, after much labor, approve, and accepted as the basis for the government of the 'American Society of Civil Engineers and Architects.' Officers were elected in accordance with the provisions of the Constitution, and the Society was duly organized.
>
> It becomes our duty, in conformity to a resolution of said Society, to address such members of the respective professions as are known to us throughout the country, and, laying before them in brief the result of their deliberation, invite them to cooperate in a furtherance of the aim and objects of the Society, so far as they may be found to accord with their individual views. Such gentlemen only as receive this circular are eligible as members of the Society by a bare notice of their desire to become such and a compliance with the accompanying forms (on or before December 1st, 1852). All others will be elected by ballot in conformity with the requirements of the Constitution.
>
> It will be admitted that no point in our country offers the facilities for rendering such a society of practical benefit to the public as well as to its own members as the city of New York, and so long as this city retains its present commercial importance, so long it will be a

> center around and within which there will accumulate by a natural law practical commercial and professional information not elsewhere to be sought. …

The Constitution of the Society declares that it has for its object:

> The professional improvement of its members, the encouragement of social intercourse among men of practical science, the advancement of engineering in its several branches, and of architecture, and the establishment of a central point of reference and union for its members.
>
> Civil, geological, mining and mechanical engineers, architects and other persons who, by profession, are interested in the advancement of science, shall be eligible as members.
>
> It is anticipated that the union of the three branches of civil and mechanical engineering and architecture will be attended by the happiest results, not with a view to the fusion of the three professions in one; but as in our country, from necessity, a member of one profession is liable at times to be called upon to practice to a greater or lesser extent in the others, and as the line between them cannot be drawn with precision … [the union will] do much to quiet the unworthy jealousies which have tended to diminish the usefulness of distinct societies formed heretofore. …
>
> In reference to the revenue of the Society … the initiation fee might, with propriety, be fixed as the same for all members; while the yearly contribution or assessment, will, in the case of the resident members, be double that of the non-resident. The rates were accordingly fixed at $10 initiation fee for all members, and $10 yearly assessment on each member residing within 50 miles of the city of New York, and $5 yearly assessment on all members residing beyond.

By the next meeting, December 1, 1852, 10 more members had responded to the invitation. At that meeting, Bylaws were read, discussed, amended, and adopted, and the formal organization of the American Society of Civil Engineers and Architects was completed.

The 10 men who paid dues on or before December 1, 1852, becoming "Charter Members" together with the 12 founders, were John F. Winslow, McRee Swift, John A. Roebling, Robert A. Brown, E. French, H.A. Gardner, Archibald Kennedy, James B. Francis, I.C. Chesborough, and George M. Dexter.

All of the 22 Charter Members were engineers except Dexter, a Boston architect. The 1852 effort may have been more successful than that in 1839 in part because the participants were few, all located in or near New York City, and probably well known to each other.

The Society held a meeting January 5, 1853, with eight members present and authorized a circular soliciting the donation of reports, maps, plans, and other data from those in charge of public works. Under the title "The Relief of Broadway," President James Laurie presented a proposal for placing railway tracks above the level of the street.

Eight meetings were held in 1853 and, while President Laurie came to each, average attendance was only six. In addition to the Elevated Railway proposal, the members discussed projects to improve Church and Mercer Streets in New York City, "The Use and Abuse of Iron as Applied to Building Purposes," and a presentation by J.W. Adams on the Lexington and Danville Railroad suspension bridge over the Kentucky River.

The first *Annual Report* of the Board of Direction, dated October 10, 1853, recorded a total membership of 55, receipts of $700, and expenditures of $115.12. The precarious state of the Society is documented by this excerpt:

> In view of the limited number of resident members, the policy of the Board during the past year has been to husband the resources of the society, to make no expenditures that could well be avoided, so that in case of failure the funds collected might be returned to the members. ... For these reasons no steps have been taken towards the formation of a library, or for renting rooms for the use of the Society.
>
> The Board regrets that they cannot speak in more flattering terms of the success of the Society, or with more confidence of its future prospects, but, believing that such an institution is much wanted, and that it rests and is entirely within the power of those eligible as members to make it eminently useful, they recommend ... that renewed efforts be made to obtain additional members who are residents of the city or vicinity, and can attend the meetings.

The first *Annual Report* also listed the roster of members (Table 2-1).

Average attendance of the six meetings in 1854 was even lower than in the year before. The *Annual Report* for that year showed but 54 members, despite the fact that dues for residents had been reduced from $10 to $5 and for nonresidents from $5 to $3. There were only 10 resident members. In reducing the dues "to render membership less onerous," the officers noted that "the labors of the members more than their money is wanted to make the Society useful."

At the 1854 meetings, members discussed the replacement of an aqueduct on the Morris Canal, materials used for water conveyance, the comparative economy of inclined planes and steam locomotives on railroads, and "Ball's indestructible water pipe."

The struggle for membership interest and participation continued into 1855, when the first formal paper was presented, "Results of Some Experiments on the Strength of Cast Iron" by W.H. Talcott. Other topics included "Blasting Rocks by a New Process," a comparison of English and American wire rope, and "Recent Inventions for Economizing Fuel in Generating Steam."

Table 2-1. *Roster of Members Listed in the First* Annual Report, *1853*

Honorary Members	George Sears Greene
John James Albert	Daniel L. Harris
Alexander Dallas Bache	Waldo Higginson
Henry Burden	George E. Hoffman
Dennis Hart Mahan	Josiah Hunt
Moncure Robinson	M.B. Inches
Joseph G. Totten	Theodore D. Judah
	Archibald W. Kennedy
Corresponding Member	James P. Kirkwood
T.S. Brown	James Laurie
	Isaiah William Penn Lewis
Members	William Jarvis McAlpine
Julius Walker Adams	John McRae
James Barnes	Thomas C. Meyer
E.L. Berthoud	J.F. Miller
Robert N. Brown	D. Mitchel Jr.
I.C. Chesbrough	James E. Montgomery
Stephen Chester	William H. Morell
Charles W. Copeland	William W. Morris
Alfred Wingate Craven	James Otis Morse
Matthias Oliver Davidson	Thomas S. O'Sullivan
George M. Dexter	William D. Picket
Thomas Addis Emmet	Simeon S. Post
James K. Ford	John Augustus Roebling
James Bicheno Francis	William H. Sidell
Edmund French	Israel Smith Jr.
Edward Gardiner	McRee Swift
Henry A. Gardner	William H. Talcott
Robert B. Gorsuch	William Wallace
H. Grassau	John F. Winslow

James O. Morse, who was elected in December 1854 to the combined offices of Secretary and Treasurer, resigned January 5, 1855, effective with the appointment of his successor. Morse held office for 12 years until 1867. By then, it was clear the Society needed both a publication to enhance communication with nonresident members and a permanent headquarters. At that time, all meetings were held at the Croton Aqueduct Department.

The need for rooms to house offices, meetings, a library, and a museum was addressed in a March 2, 1855, committee report presented to the Board of Direction:

> We have no place now where country members can find the Society except once a month, and no place where any papers, reports, or maps, which we have collected, can be seen or consulted except at the monthly meetings.

> As country members may be frequently in the city between the monthly meetings, and can but rarely suit their visits to the time of these meetings, and as we believe that a sufficient room can be obtained at an expense not inconsistent with the means of the Society ... so that those who are unable to attend the monthly meetings of the Society can yet find and have access to any information on its files.
>
> To such a room we would by advertisement invite all persons having models of patented or proposed improvements to exhibit, to send them occasionally, for the inspection of such engineers or architects as might be in the city. We would endeavor to make the room in this way a point of interest for engineers, and for all inventors or others who have anything to communicate or explain to the profession.
>
> We suggest that the experiment be tried for one year, and to that end recommend that a committee be appointed to procure a room convenient, if possible, to the office of some member who may be willing to take supervision of it, and that a desk and other necessary furniture be procured, provided that the rent of the room not exceed $250.00, and the cost of furnishing it do not exceed $150.00.

The Board considered the report and the record shows only that "no motion was made, and the Society adjourned." Charles Warren Hunt, a former Secretary and earlier historian of the Society, observes: "No report of this discussion has been found, and it is difficult to understand why the experiment of giving to the Society a local habitation was not tried, it having been clearly demonstrated during its two years of life that without one success was impossible."

The next meeting of the Society took place 12 and a half years later. The Civil War (1861–65) surely prevented further progress, although it appears that not more than a dozen of the members in 1855 were active in the war in a military capacity.

The gap in ASCE activity may have also been related to James Laurie. After a three-year study of bridges for the State of New York, he spent two years in Nova Scotia evaluating railroads and planning extensions. From 1860 to 1866, he was chief engineer of the New Haven, Hartford, and Springfield Railroad, during which time he designed and built an iron and lattice bridge to replace a wooden structure carrying a single railroad track without interruption of traffic on the line. In fact, no steps were taken to revive ASCE until James Laurie returned to New York City in 1867.

A New Beginning

President James Laurie called a special meeting of the Society, which was held on October 2, 1867, in C.W. Copeland's office. Others present were J.W. Adams, James K. Ford, W.J. McAlpine, James O. Morse (still Secretary and Treasurer), Israel Smith, McRee Swift, and W. H. Talcott. The attendees agreed unanimously "to take such steps as might be necessary to resuscitate the Society." Copeland, McAlpine, and Morse were asked to prepare "a plan for the revival of meetings ... not to call for the expenditure of more than $1,200 for the coming year."

Vice President Copeland presided over the next meeting, which was attended by Adams, McAlpine, Morse, Swift, Talcott, and James How. The committee report excerpted here was adopted and the Board of Direction authorized to implement its recommendations:

> October 8th, 1867
>
> *To the American Society of Civil Engineers and Architects:*
>
> The Committee appointed at a meeting of the Society held October 2d, for the purpose of proposing a plan for the more permanent establishment of the Society, beg leave to report:
>
> It is, we think, true, that all societies similar to ours, that have been successful and grown to greatness, have had their beginning in a small way, but in those beginnings the social element as always been cultivated, and out of the frequent and pleasant meetings, in a social way, of a few men of kindred tastes and pursuits, have grown most, if not all, of the permanent associations of the world, that, like ours, are devoted to science and to art.
>
> We have offered to us, on the corner of William and Cedar Streets, two blocks from Broadway, and two short blocks from Wall Street, two rooms in the third story, directly over the rooms of the Chamber of Commerce. The building is elegant, its entrance and stairway are commodious, and there is altogether an air of quiet respectability about the place. ... Together with these advantages, there is one which may contribute a good deal to the fitness of the place for our wants. We allude to a private restaurant in the upper story of the building, kept by the janitor, for the occupants and gentlemen of the (neighborhood) banks and insurance companies. ...
>
> The rooms are connected by sliding doors. The larger one is about 20 × 14, the other about 14 × 10; both have grates, and the larger one has fixtures for Croton water. Gas outlets are provided suitable to our wants, but there are no gas fixtures. The owners will paint and clean the rooms, and intimated that they would give us a new handsome grate and mantel. The rent is, in our opinion, remarkably low, it being but $400 for one year. We can take the rooms for a term of years, or for one year.

The Board recommended that a committee engage the rooms, furnish them at a cost not exceeding $600, and prepare additions to the Bylaws of the

Society "as may be deemed necessary to serve for the proper care and management of the rooms." The financial thread by which the Society clung to life is clear from the following memoir:

> Mr. Morse will be remembered as the first Treasurer of our Society, who, after the war, was instrumental in its revival. The Society had languished for several years, in fact, was considered as having died out entirely … when it was proposed … to organize a new society, but the lack of funds was an obstacle … nothing considered to be available from the wreck of the old organization, which was regarded as having died intestate. Mr. Morse stepped forward and presented an account in detail, showing every dollar that the Society had received and expended, and allowing compound interest for the difference, showed a very respectable balance to the credit of the Society, and upon this the reorganization was based which has resulted so successfully.

The cash balance held by the Treasurer in a savings bank had increased from $266.93 to $497.57 between 1855 and 1867. Laurie presented a check for $558.25, the dividends at compound interest he had collected as President on five shares of New York Central Railroad stock held by the Society. The total assets came to almost $1,600, more than enough to meet all needs.

A letter from Laurie to Morse on November 20, 1867, indicates that the entire funds of the Society were lost through speculation in 1854 or earlier, when the officers decided that the principal portion of funds would be invested in stocks. The New York Central stock certificate was somehow misplaced during the 1855–67 inactive period, and the company withheld payment of dividends. It was not until 1878 that the value of the original shares plus accumulated dividends, totaling $1,475.79, was recovered.

The Society's Annual Meeting on November 6, 1867, was the first to be held in the new headquarters in the Chamber of Commerce Building, 63 William Street. The revived ASCE admitted 54 new members at the next meeting, initiated regular fortnightly meetings, and began collecting library materials. After the first issue of *Transactions* a few years later, there was little doubt about the stability and promising future of the new organization.

In his presidential address at the 1870 Annual Convention, Alfred W. Craven reflected on the confidence that pervaded the Society:

> Our Society, as you all know, was first formally organized in the year 1852. … We met, and strove our best to nurse the infant in whose health and growth we all had so deep an interest. We met, as our early records show, to the number of seven, five, sometimes three on an evening, and every man present was an official. We could not afford to hire a room, and so [we] met at night in one of the vacant rooms of the Croton Aqueduct Department in this city. … Even if we did nothing else, to courageously and persistently attend such meetings

> was to work hard. The struggle lasted three years, until at last, in March, 1855 we were obliged to succumb. ... Some however, still retained their courage, and urged the careful investment of our funds and the safe storage of our books and papers until hoped-for brighter days. Our syncope lasted twelve years; and when, resolved upon one more strong effort for life, we came together in October, 1867, we started with a fund which added materially to our strength and spirits. We took our present rooms, and, so far as our limited means permitted, we tried to make them attractive. We were no longer vagrants ... so we all felt renewed hope. ... At the time of regathering for this work, the members on our old register ... numbered in all only twenty-eight. Of those who were present in the city, there were but thirteen. This was in October 1867, not yet three years ago. Now we count upon our rolls 179 regular members and the number is constantly increasing.

At the organization meeting of the Society in 1852, attendees resolved to incorporate under the name "American Society of Civil Engineers and Architects." The incorporation procedure was somehow overlooked, and this name never became legal. The American Institute of Architects was established in 1857, so it was not surprising that the members voted on March 4, 1868, to change the name to "American Society of Civil Engineers." Once more, the Society waited, and it was not until April 1877 that the name was registered legally.

No chronicle of the birth of the American Society of Civil Engineers would be complete without acknowledging the perseverance of James Laurie. He insisted that a national engineering society was necessary and feasible, and he led the discussion that led to the November 5, 1852, meeting. Laurie was the Society's first President, and he carried it through those difficult early years. Perhaps his greatest contribution was his effort toward ASCE's revival.

By 1875, the membership of the Society included six Honorary Members, 362 Members, 17 Associates, and 23 Juniors—a total of 408. About 70% of these were Non-Resident Members, residing beyond 50 miles of New York City. It was no longer true that ASCE was a "New York City gentleman's club," an apt description in the beginning.

The professional accomplishments of its leadership and members also gave stature to the Society. These engineers were dominant figures in the design, construction, and management of the major railroads, canals, roads, municipal services, and other public works projects of the day.

ASCE evolved throughout a quarter-century. The process was a part of the maturing of the engineering profession, which by this time had made countless contributions to the development of the United States. These men were so preoccupied with building a nation that they were unable to give adequate attention to their own desires and needs. This is a manifestation of the service ethic, the hallmark of the true profession.

CHAPTER 3

Money and Management

America's first national engineering society was handicapped from the start by limited resources, sometimes ponderous procedures, and complex administrative involvements. In the eyes of some, these handicaps persist today. The manpower and fiscal resources of the Society—almost entirely contributed by its members—were indeed modest compared with its goals.

The management structure of the Society was prescribed in its Constitution, supervised by the Board of Direction, and executed by the staff. For many years, members did virtually all the actual work of the Society. After about 1940, ASCE expanded its staff and retained outside specialists.

This chapter reviews the evolution of the Constitution and related rules, the Board of Direction, the staff, the organizational structure (national, regional, and local), the headquarters facility, the finances of the Society, and the most important resource of all: the membership.

The Constitution and Bylaws

Discussion of the Constitution and Bylaws has engaged more time in the business meetings of the Society and its Board of Direction than any other single topic. The Constitution was amended 44 times between 1852 and 1974. More than half of the amendments dealt with membership requirements or admission procedures, or with procedures for nomination and election of officers.

The Society overhauled the Constitution in 1891, 1921, and 1950, and it enacted significant amendments dealing with the composition and authority of the Board of Direction in 1897, 1908, 1930, 1966, and 1970. Amendments establishing geographical representation districts (1894) and extending full membership privileges to Juniors (1947 and 1950) were also important.

Proposed constitutional amendments were sometimes controversial. In 1921, differences of opinion were so bitter that the Board ordered a portion of the record expunged. At issue was the end to ex officio membership of the Secretary and Treasurer on the Board of Direction.

The first Bylaws were a combination of rules of order for meetings and elementary operational regulations. Amendments were so prolific as to defy detailed reporting. With the 1921 redraft of the Constitution, the Bylaws became essential to the administration of the increasingly complex membership and election procedures and the growing number of committees. This trend continued, and in October 1951, a third level of regulation was created in the Rules of Policy and Procedure that supplemented the Bylaws. By 1974, amendment of the Bylaws and the Rules was invariably a part of every Board of Direction meeting.

There is no apparent reason for the concern with procedural detail throughout the life of the Society. This propensity has drawn on the time and stamina of Board members, and it sometimes has diverted attention from business more relevant to Society objectives.

The Board of Direction

From 1891 to 1897, when all Past-Presidents served, the size of the Board varied from 35 to 39 members. Initially, the Board's authority was limited to "general care of the affairs of the Society" and management of its funds. All major decisions were made by vote of the membership, either in general business meeting or by letter ballot. Special committees could be authorized only by majority vote of the Society. Membership admissions were subject to letter ballot of the Society, with three negative votes excluding.

In 1908, the Board was empowered to elect members and to transfer them in grade. Special committees could be appointed by the Board only when authorized by a general business meeting until 1921.

In 1970, the Board of Direction transferred the schedules for annual dues and entrance fees from the Constitution, which could be amended only by membership referendum, to the Bylaws, which could be amended by Board action alone. This action gave the Board complete management authority over the financial affairs of the Society.

In 1964, about 29% of ASCE members were engaged in private practice, 40% in public practice, 7% in education, and 24% in all other areas. Representation on the Board of Direction has not followed this pattern, although it has been reasonably compatible with the geographical distribution of the membership. Private practice has always been strongly represented on the Board, and education is gradually assuming a stronger role in management. The declining role of the railroad executive is evident. In 1900, 2% of members were railroad executives; by 1950, this had dwindled to 1%; and by 1974, there were no members reported in this category.

The public practice sector has consistently been underrepresented, probably because working conditions in government are not always conducive to outside activity. More liberal policies on time for professional activities, travel reimbursement for attendance of meetings, and recognition of professional development might correct this imbalance.

The importance of Society leadership is evident in the procedures for nominating and electing officers. The President, Vice Presidents, and Directors served one-year terms from 1852 to 1891; thereafter, the President served for one year, the Vice Presidents for two years, and the Directors for three years. The Secretary and Treasurer were elected annually by the membership until 1894, and after that by the Board; both were members of the Board until 1921.

From 1852 to 1891, all officers were elected by the membership at the Annual Meeting; a small Nominating Committee was set up in 1878 to expedite the process. The new Constitution in 1891 called for a 19-member Nominating Committee with election by letter ballot of all Corporate Members. This system prevailed until 1921. When the Districts were first created in 1895, it was specified that the Board would include a Vice President and six Directors representing District 1 (the New York City area).

In 1921, the Nominating Committee was discontinued and all officers were nominated by a two-ballot system, but six years later a Nominating Committee comprising the Directors and Past-Presidents was authorized to designate the "Official Nominee" for the office of President. The Society adopted a single nomination ballot in 1950 to determine official nominees for the Vice President and Director.

In 1966, the Board approved the automatic advancement of a President-Elect into the presidency after a year in the post. This ensured familiarity with current Society affairs by the new President.

The nomination of Vice Presidents and Directors by mail ballot of the membership, never a popular process, became increasingly awkward and expensive as the Society grew. These procedures were under review in 1974, and return to one or more nominating committees appeared to be imminent.

Because the Society was organized in New York City, there was sensitivity to domination by "New York" as ASCE grew nationally. Until as late as 1921, the constitution required that one Vice President and six of the 18 Directors on the Board of Direction be elected from District 1. Since 1921, the number of Directors has been based upon membership distribution, and the representation from District 1 gradually diminished to two of 19 Directors in 1974. Nevertheless, when the fund-raising campaign for the United Engineering Center was under way in 1958, there was still some feeling against the "ghost" of the New York hierarchy.

Actually, the influence of New York City in the Board of Direction was declining, and a new concentration of power was growing in California. In

1974, the four California Sections were included in District 11, which was allocated four of the 19 Directors on the Board.

The Headquarter Offices

The Society's first home was a two-room suite in the Chamber of Commerce Building, 63 William Street, New York City, and the first Annual Meeting was held there on November 6, 1867. The Society took an additional room four years later, then needed more space by 1875, and rented new rooms at 4 East 23rd Street. By 1877, ASCE headquarters were in a house at 104 East 20th Street, but these quarters also proved unsatisfactory. On January 1, 1881, a letter to the membership stated that the Society needed a "more permanent home, properly arranged for its use," and sought contributions for a building fund. The goal was $25,000 to $30,000.

In April 1881, ASCE bought property at 127 East 23rd Street for $30,000, with a $5,000 down payment. This sum was raised by $500 advances on the part of 10 members, to be repaid from subsequent subscriptions to the building fund. This house accommodated the library and the Society operations until 1897, by which time it had been improved to a value of $50,000.

ASCE's growth led to an ambitious project in 1895, when the Society bought two lots at 218–220 West 57th Street for $80,000. Constructing a suitable building on the site would cost $90,000, so ASCE had to raise $60,000 by building fund subscriptions and then took on a $60,000 mortgage to cover the total cost of $170,000.

Board action required that "the construction and architecture of the new Society House be entrusted to Members of the American Society of Civil Engineers and to none others," and plans were prepared by Joseph M. Wilson, ASCE Vice President (1894–95). At the 1896 Annual Meeting, the Board decided instead to hold a design competition, open to a selected list of architects who were not members of ASCE. C.L.W. Eidlitz was the successful candidate.

The ground floor of the new "House" had a large foyer, three office rooms, and a spacious lounge for informal and social gatherings. The second floor had a reading room and 500-seat auditorium. The third floor was devoted entirely to office space and the fourth to library storage for 150, 000 volumes. The building facade was designated as a 57th Street landmark years later. (Figure 3-1)

The building formally opened on November 24, 1897. The Society bought additional land and a 50% expansion of the building was constructed in 1905, bringing the total investment in the property to $360,000.

From the beginning, members were disappointed in the acoustics of the new auditorium. In 1909, a special committee recommended carpeting, certain wall hangings, and other treatments in hopes of improving the acoustics.

Figure 3-1. *The Society's 57th Street Building in New York, 1897 (Courtesy Avery Library, Columbia University)*

The problem was still unsolved 13 years later, when a long-suffering member made a stirring appeal for relief:

> I think it is a reproach to a Society as scientific as ours … that it continues to maintain an auditorium with such uncommonly poor acoustic properties. … I remember the first meeting we had in this room. … It was soon perceived that members were changing their seats to hear what was being said … and somebody asked whether something could not be done to improve these acoustic defects. Some member suggested that wire might be hung across the room, and one man misunderstood the meaning of the suggestion and approved it on the theory that it was a motion to hang the architect!

This plea elicited action to "employ the best talent available" to remedy the situation. The architect escaped unscathed!

In the midst of this building project, this startling letter appeared:

> 2 East 91st Street
> New York, February 14, 1903
>
> Gentlemen of the
> American Society of Civil Engineers
> American Society of Mechanical Engineers
> American Institute of Mining Engineers
> American Institute of Electrical Engineers
> and the Engineers' Club:

Figure 3-2. *West 39th Street Headquarters Building*

It will give me great pleasure to give, say, one million dollars to erect a suitable Union Building for you all, as the same may be needed.

With best wishes,
Truly yours,

Andrew Carnegie

In a mandated membership referendum, ASCE decided not to participate by a vote of 1,139 to 662. The Society then withdrew from cotrusteeship of the Carnegie gift. ASME, AIME, and AIEE proceeded to form the United Engineering Society to administer conversion of the Carnegie gift into the Engineering Societies Building at 33 West 39th Street, New York City (Figure 3-2).

ASCE, unlike AIME, ASME, and AIEE, was housed in a handsome new, well-financed headquarters building. The Society might have opted to accept the Carnegie gift, but some members seem to have thought that a merger of the societies was intended rather than a venture to own and manage a common building. Those members saw the identity, autonomy, and high membership standards of ASCE as at risk. There was particular reluctance to accept affiliation with the New York Engineers' Club, a private social organization that was included in the Carnegie offer. Such ASCE notables as Octave Chanute, Oberlin Smith, Rudolph Hering, and Charles Macdonald favored participation, but the proposal appears not to have been clearly understood by the membership.

By 1914, the Engineering Societies Building funded by Carnegie was cleared of debt, but many felt the venture was incomplete without the oldest of the national societies, ASCE. In 1915, the United Engineering Society offered ASCE full status as a Founder Society for $250,000. Most of that money covered the cost of adding two floors to the new building for ASCE.

This time, ASCE's members voted 2,500 to 300 in favor of affiliation with the United Engineering Societies—a strong endorsement of the implicit engineering society unity. The Engineering Societies Building was enlarged by two floors. The additional loads had to be carried to the foundation by an independent structure, and the actual cost of the work came to $262,500. On December 17, 1917, ASCE became an occupant and the fourth Founder with appropriate pomp and fanfare.

The Society then leased the 57th Street property, a highly profitable investment. In 1966, ASCE sold the property for $850,000. The Society's expenditure, including the land and the building with its 1905 addition, was $366,000.

The Building needed more space by 1950, and the Societies undertook a study of larger quarters. The study led in 1957 to acquiring land on First Avenue facing the United Nations complex. The plan was to build a 20-story tower at a cost of $12.8 million. The engineering societies joined forces to raise $3,787,000 in contributions from their members, with quotas for each.

Under the slogan, "Let's Get the Job Done," ASCE raised its quota of $800,000 in just 33 months, beaten only by the American Institute of Chemical Engineers, with a smaller quota of $300,000 and a membership concentrated in a few very large corporations. Success was largely a result of efforts by ASCE's Local Sections. The first five to reach their quotas were the Kentucky, Lehigh Valley, Nashville, Cincinnati, and Columbia Sections, in that order.

Simply by being the first Society to ask the United Engineering Trustees for the space, ASCE was assigned the top two floors of the new building and moved in September 5, 1961. In 1974, staff growth required a third floor. At that time, some of the other societies were becoming restive because of high costs and a shortage of competent office personnel in New York City. A few moved out of the building, a concern to the building's Trustees.

The United Engineering Center proved an efficient and functional headquarters, despite the urban problems plaguing New York City. The location in the prestigious United Nations complex gave a measure of status to the engineering profession. As long as the property was exempted from New York City taxes, the economic factor would be favorable. Unless there was some major complication, it seemed likely in 1974 that ASCE would support continuation of the center in its 47th Street location (see Figure 3-3).

The Library

Among ASCE's first actions was the issuance, in January 1853, of a circular to "all men in charge of public works, asking for printed reports, maps, plans,

Figure 3-3. *United Engineering Center, 345 East 47th Street, New York*

etc., in order to start an Engineering Library in connection with the Society." There were no funds for acquisitions, but members made some generous donations. Some were gifts of extensive personal collections, such as those by William Y. Arthur, M.ASCE, in 1872 and by Past-President William J. McAlpine in 1873. By 1873, the library contained 3,433 items, and a newly created committee drafted library policy. An 1885 attempt to establish a joint library for ASCE, AIME, ASME, and AIEE was one of the earliest efforts toward intersociety cooperation. The effort was abandoned three years later.

ASCE's collection had grown to 22,000 items by 1897 when the Society moved into its new 57th Street building. The library merged with those of the other Societies in October 1916, and became known as the Engineering Societies Library (ESL). At that time, there were more than 89,000 accessions. ASCE presented 2,000 duplicate volumes to the Cleveland Association of Members. In the years to follow, the ESL became one of the outstanding engineering libraries in the world. Funding depended, however, on a per capita assessment against the constituent bodies of United Engineering Societies. During the "Information Explosion" of the 1960s, the need for information storage and retrieval exceeded the resources of the ESL, and some supporting societies were reluctant to continue the membership assessment.

The Staff

Until 1872, the Society staff was the Secretary alone, elected by the membership, who served without compensation. Five incumbents during that period served but one year, the exception being James O. Morse, who resigned on

January 5, 1855, but was not replaced until 1869, following the Society's 12-year hiatus. Morse also served as Treasurer for 21 years.

In June 1872, it was "the unanimous sense of the membership ... that it is injurious to the dignity of the Society to longer accept the services of any gentleman acting as permanent Secretary without compensation." Gabriel Leverich became the first paid Secretary, being required "to devote all time necessary to the thorough development of the Society's interests" for an annual salary of $3,000. He was empowered to hire any necessary assistants, but they had to be paid from his own salary. Duties of the Secretary included correspondence, meetings, membership development, responsibility for professional functions, and service as Librarian.

This arrangement continued until 1892, except for relaxation in 1885 of the requirement that the Secretary pay his assistants from his $3,000 salary. The new Constitution in 1891 provided Secretary Francis Collingwood with an Assistant Secretary, Charles Warren Hunt, who took this position and also became Librarian. Hunt later served the Society as Secretary for a quarter-century and became its first official historian.

ASCE's staff in 1892 consisted of the Secretary, Assistant Secretary, Auditor, Assistant Librarian, one clerk, and two stenographers—a total of seven full-time employees. A janitor and two office boys were also on staff, along with a part-time Treasurer and one stenographer. The total payroll was $12,722. By 1900, salaries totaled $16,283, with nine full-time staff in addition to janitors, office boys, and part-time help. When the Society moved into the Engineering Societies Building in 1916, there were 22 staff members exclusive of maintenance personnel. Staff operations were administration, publications, meetings, and the library.

The 1930 Functional Expansion Program revolutionized the range of Society activity, and for the first time staff members were assigned to full-time service in professional development functions. ASCE's records yield few details, but in 1940 at least 4 of the 56 staff members were engaged with public affairs, education, employment conditions, and other professional activities.

Staff decentralization was emphasized during this period. In 1935, ASCE created a staff post of Field Secretary "to bring to Local Sections a view of their new responsibilities and opportunities." The Washington (Eastern) Field Office was opened mainly as a wartime facility in 1941, but it continued into the postwar years. ASCE authorized a West Coast Field Secretary in Los Angeles in 1944, and a Mid-West Regional Office in Chicago followed two years later. Apparently the benefit-cost ratios were unfavorable, and the Chicago and Los Angeles operations were closed out at the end of 1948. The Washington office was suspended in 1955, then reestablished in 1972.

In 1950, the Society employed 72 people, 50 of whom worked on general member services, 21 in publications, and 6 in professional development. In 1974, the staff grew to 109 employees, with 52 in general services, 38 in publications, and 19 in professional development.

From 1852 until 1972, the staff was headed by the following individuals with the various titles of Secretary, Secretary and Librarian, and Secretary and Executive Director: Robert B. Gorsuch, 1852–53; Edward Gardiner, 1853–54; James O. Morse, 1854–69; Thomas C. Meyer, 1869–70; Alfred P. Boller, 1870–71; Gabriel Leverich, 1871–77; John Bogart, 1877–91; Francis Collingwood, 1891–95; Charles W. Hunt, 1895–1920; Herbert S. Crocker, 1920–21; John H. Dunlap, 1922–24; George T. Seabury, 1925–45; William N. Carey, 1945–55; and William H. Wisely, 1955–72. Eugene Zwoyer became Secretary and Executive Director in 1972.

Two Executive Directors, John H. Dunlap and George T. Seabury, died in office. Charles Warren Hunt served for 25 years, the longest of any.

In 1974, there were 1.5 staff members per 1,000 members of the Society, a figure well below the ratio in the other Founder Societies. ASCE nonetheless conducted a broader range of professional activity and maintained a comparable level of membership services in other areas.

Public Recognition

Concern about national recognition for the civil engineering profession dates back to 1874. The President and Vice Presidents at that time studied the feasibility of a federal charter as a means of gaining such stature, and they recommended that an application be made to Congress "for a national charter, upon the most favorable terms which may consistently be granted, keeping in view mutual advantages between the Society and the country." No action is reported.

Almost 50 years later, in 1922, Past-President Clemens Herschel again urged that the Society seek such a federal charter. This movement foundered when a complicated opinion by legal counsel concluded:

> Assuming that the Federal corporation is authorized to operate with the same powers as those possessed by the existing corporation, it will have no advantages of power, privileges, or immunities over those possessed by the ASCE. No one can say whether Congress would or would not confer special or extraordinary privileges over the new corporation, but presumably it would not.

As early as 1873, there was restiveness on the part of Non-Resident Members concerning the concentration of Society meetings and management in New York City. Considerable attention was given to this situation at the Louisville Convention that year, and something of a crisis was averted by extension of the voting privilege to Non-Resident Members by letter ballot.

Ten years later, in 1883, a failed constitutional amendment would have authorized Sections "for the advancement of a special Branch of Engineering." There was concern at the time that a new "Society of River and Harbor

Engineers" would be formed if ASCE did not act. (The event preceded by more than 100 years the ASCE organizational move to Institutes in the late 1990s.)

At the 1885 Annual Convention, Arthur M. Wellington, M.ASCE, called for a task committee "to consider the matter of making such changes in the organization of the Society as may be desirable in connection with the subject of local engineering societies or clubs, and of sections or chapters of the Society; also to take into consideration the future policy of the Society in relation to the admission of branches of engineering." Only the Western Society of Engineers (Chicago) and the Civil Engineers Club of Cleveland were mentioned as local bodies. The committee was formed, then discharged a year later when its members failed to reach agreement.

The Board of Direction then invited comments from members and from local engineering clubs on the subject of national organization of the engineering profession. Only 11 members responded, in addition to the Engineers Club of Saint Louis, the Boston Society of Civil Engineers, and the Western Society of Engineers.

In April 1905, associations of ASCE members similar to that in Saint Louis were formed in Kansas City and in San Francisco. The views expressed were so diverse that the Board concluded, in 1887, "that there is at present no desire for changes in the organization."

Nevertheless, on February 29, 1888, 16 ASCE members became the Saint Louis Association of Members, " to debate the merits of local projects." This was the first local organization of ASCE members that was not officially a part of the Society until 1914.

A move in 1890-91 to codify and update the ASCE Constitution included efforts to develop the local identity of the Society. A narrowly defeated provision called for merging with the 27 then-existing local clubs and associations. At the Annual Convention, the Board noted these groups and recommended that "wherever possible, steps be taken calling attention of the membership to (local organizations), and that such Associations be formed." The Society also adopted a set of Bylaws for proposed local associations.

By 1915, there were 17 Associations of Members: Kansas City, San Francisco, Memphis, Colorado, Atlanta, Philadelphia, Seattle, Portland, Los Angeles, Texas, Spokane, Louisiana, Baltimore, Saint Louis, Northwestern, Cleveland, and San Diego, in order of their formation.

At the 1915 Annual Meeting, the presidents of 14 of these associations asked for a vastly increased say in Society affairs, recommending that

> the Society be divided geographically into Districts; every member residing in the territory covered by that District to become a part of the District Organization without payment of further dues. Each District to have a President, or Chairman, and Secretary, and existing Local Associations in that District to become Local Sections reporting to the management of the Society through the District Organiza-

tion; each District to elect its own representative on the Board of Direction of the Society.

ASCE held a general meeting April 19, 1916, with representatives of AIME, ASME, and AIEE, to explore the subject, "What relations should exist between the National Engineering Societies and the local sections or associations of their members, and, in the interests of the Profession, what should be the attitude of both of the above to other local engineering societies or clubs?" The outcome was a simple plan to advance the "Solidarity of the Engineering Profession" through a joint National Conference Committee and a cadre of joint Local Conference Committees, set forth by ASCE President Elmer L. Corthell. The plan was not adopted.

An earlier step toward a regional structure had been taken in 1894, when seven Districts were designated simply as a means of achieving a measure of geographical representation on the Board of Direction. A change from seven to 13 of these representation Districts in 1915 was the only regional development resulting from the recommendation of the Conference of Local Association Presidents earlier that year.

In April 1917, the task committee studying the matter reported to the Board, in part as follows:

> The Local Associations of the American Society of Civil Engineers are in many cases responding to a sentiment that there should be more cooperation with other Engineering Organizations, more influence exerted in local communities.

The 1915 conference of presidents of 14 Associations approved the plan of district representation, but when the Secretary asked for an expression of opinion, only four Associations responded, and only one of these advocated the Districts plan. The Associations that offered suggestions agreed they wanted few restrictions and freedom to deal with local affairs.

These views prefaced a thorough statement of policies and rules on relationships of the local units to the national Society, to other engineering organizations, and to the public. The report was adopted, and it became the guideline for further organization of local Associations of Members, to a total of 23 in 1920.

The (February) 1921 *Annual Register* for the first time listed "Local Sections" instead of Associations of Members, even though a new Constitution was not adopted until October of that year. That document gave formal and official status to Local Sections. There were 30 such sections in 1921; by 1974 , there were 79.

The prestigious Boston Society of Civil Engineers, organized in 1848, considered merging with ASCE's Massachusetts Section for 50 years, and finally did so in April 1974. The delay was due to the Boston Society's wish to preserve its identity and traditions.

From the standpoint of action in public affairs at the local level, the Board favored Branches at the state level. The original constitutional authorization permitted Sections "in any locality." The Texas Section, with 13 Branches, was organized in 1974. Some states, such as Ohio and New York, each with six Sections, set up State Councils to coordinate actions. A trend toward regional structuring reappeared in 1948 when the Sections in District 9 and in the Pacific Southwest area formed Councils primarily to expedite the nomination of Directors and Vice Presidents of the Society. There were 14 such Councils in 1974. The meetings of most Councils were limited to delegates of the Sections to take action in Society affairs, but several also offered professional programs.

There has never been a successful national move to formalize cooperation among local units of the various engineering societies, although this happened independently in several areas. ASCE's participation in a Founder Society movement toward this end was authorized in 1924, and the Society adopted a policy statement favoring coordinating Local Section functions.

In 1931, the AIEE proposed that the four Founder Societies explore the feasibility of joint state engineering councils, which would represent the engineering profession "in legislation and other nontechnical matters affecting the status of the profession and the welfare of the public." ASCE promptly endorsed and supported the venture, but it failed to survive, apparently for lack of interest on the part of AIME and ASME.

The Illinois Engineering Council coordinated engineering action at the state level for at least 35 years, as did a state legislative council in California. Issues relating to engineering registration laws in the mid-1960s led ASCE to bring the Consulting Engineers Council and the National Society of Professional Engineers together to set up tripartite "Joint Engineering Action Groups" at the state level. Several such groups were formed, and at least one of them—in Tennessee—was still active in 1974.

Since 1959, formation of new Sections has been restricted to the United States. Overseas units existed in Australia and West Pakistan in 1974.

In 1924, the Board of Direction considered creating a Women's Auxiliary after this had been done by ASME and AIME, but it concluded that such action was best left to the Sections. Authorization to set up these subsidiary bodies was formalized in the Bylaws in 1950. In 1931, however, the Tacoma Section formed its "Wives of ASCE" club, which was followed by similar groups in the Dallas and Fort Worth Branches of the Texas Section in 1938 and 1940. Other early clubs set up in the 1940s were in Cleveland, Seattle, and Indiana. In 1974 there were 16 Wives' Clubs.

The shift toward sectionalization and localization of the Society, which began in the early 1900s, reflected a similar trend nationally. The effect enhanced the Society's goal of affording public identification to the profession. This localization gained momentum, and the tightly reined administrative control from New York that had prevailed for the first half-century waned. ASCE lost some of its strict conservatism and considered national affairs.

Fiscal Matters

The astuteness of Secretary-Treasurer James O. Morse enhanced ASCE's financial strength even during the interlude of 1855–67. When the Society resumed active status in October 1867, the coffers held $1,592. Through careful management, the net worth of the Society reached $30,554 in 1875, more than a third of which was invested. Laxity in dues collection led to delayed payment of some bills in 1878, jeopardizing ASCE's credit. This led to the rigid policy on dropping members in arrears that has prevailed ever since 1879. In 1880, the Finance Committee declared "that this is the most prosperous year of the existence of this organization."

The changes in major income and expense operations of the Society are summarized in Table 3-1. The quarter-century reporting interval does not reveal several periods of financial stress in 1877, 1920, 1932–39, 1947–51, 1970, and 1973.

By far the major source of revenue has been membership dues, and the several dues increases through the years have been prefaced by strained or deficit budgets. Rapidly rising publication costs brought problems in 1920, when a $200,000 mortgage on the 57th Street property was outstanding. The Great Depression of the 1930s severely tested the Society's fiscal resources. Major assets were the 57th Street property (a $30,000 mortgage still outstanding) and the library. Liquid assets were less than $10,000 when the membership peaked at 15,190 in 1931 and then began to decline. The Society remitted more than a quarter-million dollars in dues for unemployed members, and it did not recover membership losses until 1939. During that time, curtailed expenditures together with the $30,000 annual rental income from the 57th Street building saved the day.

In 1947, publication costs increased again, along with those of expanded Society activities. A proposal that year to increase dues was rejected by the membership, requiring strict economy and curtailment of services. In 1951, dues were raised. Greatly expanded programs, particularly in the professional area, resulted in substantial dues increases in 1971 and 1973.

Until 1946, all publications were furnished to members without extra charge. That year, *Transactions* was made available by subscription; in 1962, this was done for the *Directory*, and in 1966 a subscription schedule was established for *Proceedings*. Except for *Transactions*, the subscription rates for members did not cover costs; the *Directory* and *Proceedings* were still heavily subsidized from other sources of income in 1974.

When the regular publication of *Proceedings* was authorized in 1873, it included "select advertisements, to be approved by the Committee on the Library, be received and published therewith." Income from this source was nominal, only about $2,000 annually, when advertising in *Proceedings* was discontinued after the May 1903 issue.

In 1920, the Publications Committee "was authorized to increase the page size of *Proceedings* to 9 by 12 in. and to insert proper advertising matter at

Table 3-1. *ASCE Financial Operations, 1853–1973 (dollars)*

	Income				
Year	*Fees and Dues*	*Publications*	*Advertising*	*Other*	*Total*
1853	700	—	—	—	700
1875	7,858	233	605	70	8,765
1900	42,888	2,394	2,221	4,037	51,540
1925	242,448	6,916	—	52,546	301,910
1950	463,111	66,580	148,254	93,046	770,991
1973	2,368,120	988,486	782,101	404,732	4,543,439

	Expenditures				
Year	*General Services*	*Publications*	*Technical Activities*	*Professional Activities*	*Total*
1853	115	—	—	—	115
1875	6,216	3,112	—	—	9,328
1900	27,476	17,863	—	—	45,339
1925	169,203	55,265	10,498[a]	7,841[a]	242,807
1950	373,942	313,188	20,987	31,624	739,736
1973	1,867,027	2,009,890	303,363	258,316	4,438,596

[a]Estimate.

such time as in its opinion this change would be advisable." This authority was never exercised.

Advertising income was extremely important to *Civil Engineering*, begun in 1930. It was not until 1955, when advertising sales became more aggressive, that such revenues were sufficient to cover all production costs of the periodical. The Society successfully resisted a 1971 move by the Internal Revenue Service to tax income from *Civil Engineering*.

After administration expense, publications have always taken the major share of ASCE funds. By 1925, a significant part of the fiscal resource was being devoted to professional development, and this trend has continued.

From 1881 to about 1936, the Society's reserve funds were largely committed to real estate. Reserves were sufficient to allow ASCE to waive dues for those in active military service during both world wars, as well as for unemployed members during the Depression. Surpluses of $734,317 for 1950 and $2,285,829 for 1974 included liquid assets to meet reasonable emergencies.

The Membership: The Greatest Resource

ASCE staff was minimal until 1930, and practically all the real work of the Society before that time—and much since then—was performed by the membership. At first, ASCE relied almost entirely on ad hoc committees to assume specific assignments, but gradually a complex framework of interlocking committees developed. In 1974, ASCE had about 400 technical committees, plus more than 100 administrative, professional, and awards committees and task forces.

In 1930, about 3,750 members were actively serving the Society as officers, as committee members, or in other capacities at the national and local levels. By 1970, there were more than 10,000 such members. A conscientious Local Section president might spend an estimated month on ASCE duties during his year of office, a Director three months per year, and a President of the Society half time or more. The dollar value of these services in 1974 would have dwarfed the annual budget. The demands on the time and energy of the President had become so great by 1973 that the Society studied the possibility of compensation for this full-time service, but it rejected the proposal in 1974.

Whether motivated by altruism or personal aggrandizement, or both, the working member of the organization justifies its raison d'être. ASCE has been blessed by the caliber, generosity, and loyalty of the talent available to perform its work.

The Society Badge

ASCE first considered a badge to identify members in 1882 when Director George W. Dresser obtained some designs but apparently did not submit them.

Two years later, Captain O. E. Michaelis, M.ASCE, proposed at the Annual Meeting that a suitable badge be prepared "to be worn by members at meetings, and which may be worn by them at other times." His motion was prefaced by the following:

> We were on a special train … running down to Niagara Falls, and the Secretary was furnished with a number of invitations to a collation there for distribution to the members of the Society. … It is very awkward in handing these tickets out to have a gentleman say: "Give me one of those tickets," and to have to say to him "Are you a member of the Society?" … It occurred to me that it might be appropriate … to have some modest badge … to distinguish members of the Society.

The committee, appointed in January 1884, submitted a design that was adopted by the Board and authorized for immediate sale to the membership (Figure 3-4).

Negative membership reaction came immediately. Members did buy more than 300 of the new badges, but 93 members requested that the badge design be placed before the membership. In 1892, an Annual Convention resolution requested the Board to consider new designs. A membership opinion survey in 1893 showed that a majority of respondents (303 to 214) favored a new design. One member expressed the opinion of the badge: "Its supreme merit lies in its perfect and satisfactory ugliness. It is also redolent with the prolonged perspiration of hard work in the woods, and it seems to lift our noble profession only high enough to declare us competent to survey an old farm."

At the 1894 Annual Meeting, the task committee recommended adopting the simple shield carrying the name of the Society and its founding date. The present badge design was adopted by the membership in a general business meeting. A few months later, the Board decreed

> that each member receiving a badge be required to sign an agreement to return it to the Society, receiving a credit equal to its intrinsic value, should his connection with the Society cease from any cause other than death, and waiving his right to resignation until such return.

Only Members, Associates, and Fellows were eligible to wear the badge in 1894. In 1909, the Society restricted the blue badge to Members and Associate Members, and it approved the same design for Associates and Fellows.

In 1923, ASCE approved a badge for Juniors, a blue disk with the official emblem in blue. In 1925, this badge, in maroon, was authorized for Student Chapter members. Various modifications distinguished Corporate Members from other grades. In 1974, the original blue shield designated Fellows, Members, and Affiliates; for Associate Members, the badge had a white border surrounding a blue shield; and the student badge had the same design in maroon.

Figure 3-4. *Top: Original and Present Official Badges*
Bottom: Design Showing Geometry of Official Badge

In 1949, the Board authorized a miniature lapel pin to accommodate changing clothing styles. This was replaced in 1963 by a smaller lapel-size official badge. In the late 1960s, special adaptations of the official shield identified Past-Presidents and Honorary Members.

For many years, the badges were numbered and inscribed with the holder's name, which led to some unusual experiences. In 1912, a badge still attached to the remnants of a vest was taken from the stomach of a shark caught near Catalina Island. In another instance, a member's wife gave a suit with the badge to a tramp.

The practice of membership referenda for all major decision making created a deep-seated conservatism that persisted for a century. This diminished as the Board of Direction assumed more and more management authority. The Society maintained financial independence, operating within its income. The greatest resource of ASCE has always been its dedicated membership.

CHAPTER 4

The New Profession Evolves

Any true profession prescribes and maintains standards for the education and competence of its members, and the manner in which those members practice the profession.

The educational standards must impose an obligation on the individual practitioner to remain technically current through self-study and advanced education. The standards of competence are inherent in professional membership requirements and in the legal requirements of registration and certification boards. The standards of practice are implemented through codes of ethics or other rules governing the practitioner's conduct in relationship to clients, employers, professional associates, and the public.

Equally important obligations of a profession are to provide a mechanism for collective service in the public interest and to advance the welfare of members to ensure that they shall not be inhibited by economic, political, or social constraints in the "pursuit of their learned art."

These principles of professional obligation were not clearly identified by ASCE for many years. The constitutional purpose of the Society was interpreted, in the words of President-Elect William J. McAlpine in 1868, as the advancement of "knowledge, science and practical skill among its members, by an interchange of thoughts, studies and experiences." As time passed, extensive debate about the Society's nontechnical functions led to policies addressing such activities.

The public interest as a professional responsibility of the civil engineer and of ASCE first focused on technical issues. Examples were the ad hoc committees to study "interoceanic communication between the waters of the Atlantic and Pacific" (1870), the "means of averting bridge accidents" (1873), and the tragic failures of several dams (1874).

One controversial study of the need for rapid-transit facilities in New York City in 1874 resulted in the first appraisal of ASCE policy in an 1875 report, "On Policy of the Society." The study recommended appointing "special committees on engineering subjects." ASCE expected such committees to address general aspect of an issue, without setting forth specific engineering plans.

This effort to clarify ASCE policy did not address responsibility for standards of education, competence, professional practice, and public service that denote a profession. Had this been done, it would not have taken from 1873 to 1907 to initiate an education program, from 1893 to 1914 to adopt a code of ethics, and from 1897 to 1910 to assume a leadership role in engineering registration. These developments were due to pressures from the membership.

The voice of the membership swelled in volume and authority as the number of local associations, begun at St. Louis in 1888, grew to 17 by 1915. A union-like activity in Chicago evolved into the American Association of Engineers (AAE) in 1915. The AAE—composed of a majority of civil engineers (many not in ASCE)—aimed to improve the status and economic welfare of engineers, and its emergence was observed with some skepticism by the ASCE leadership.

The Cleveland Association of Members proposed a joint Association Constitutional Conference, and the Society authorized a Committee on Development in June 1918. The resolution creating the committee included:

> Sociological and economic conditions are in a state of flux ... affecting deeply the profession of engineering in its services to society, in its varied relationships to communities and nations, and in its internal organization. ... A broad survey of the functions and purposes of the American Society of Civil Engineers is needed ... so that the Society may take its proper place in the larger sphere of influence and usefulness now opening to the profession.

The committee was chaired by Onward Bates, M.ASCE, whose given name was most apt for the job. The committee included six other members-at-large in addition to Chairman Bates plus one delegate from each of the 21 local associations then extant. What may be the most far-reaching committee report in the long life of ASCE was presented to the Board of Direction in October 1919. The recommendations included the revolutionary proposals that:

- Government of the Society be centered in District organizations of the local associations
- Membership grades be redefined.
- Activities of students and young engineers be encouraged.
- Greater emphasis be given to technical activities, research, and standards.
- New emphasis be given to such functions in the "public affairs" sector as ethics; engineering education; licensing; arbitration and expert testimony; publicity; legislation; natural resource development; service to the com-

munity, state, and nation; patent law; coordination of government activities; and industrial affairs.

- Implementation of these aims to be accomplished by a comprehensive engineering unity organization encompassing the local, state, and national levels.

A joint Conference Committee, including ASCE, AIME, ASME, and AIEE, studied an integrated approach to the resolution of similar objectives under study in all those societies. A motion to submit a questionnaire to the membership was seconded by J.C. Ralston, M.ASCE, who revealed his impatience with parliamentary protocol:

> We see scores of young engineers trooping by the doors of our respective chapter, with sneers on their faces, rushing into the meretricious arms of the American Association of Engineers. ... We ... urge ... that the American Society of Civil Engineers ... rise, if you please, above the inertia of pre-war inactivity to the mountain tops of post-war accomplishment.

The results of the mail ballot strongly favored broadened concern with "economic, industrial and civic affairs" and intersociety cooperation, but rejected the form of unity organization proposed. Also approved were recommendations dealing with technical activities, Local Sections, and the nomination and election of officers; but a suggested $5 to $10 increase in dues—to finance the proposed expansion of internal and external activities—was not popular.

The report was not implemented at once, but it was strongly manifest in the new ASCE Constitution adopted in 1921. The provisions reconstituting the Board of Direction—authorizing Local Sections and Student Chapters, revising nominating procedures, and expediting the appointment of special committees—were all related to the report. Another important spinoff was the 1925 Committee on Aims and Activities. Designated as "a systematized program to enhance the status of the civil engineer in the minds of his public," the Committee's Functional Expansion Plan created three new departments: Technical, Administrative, and Professional.

The Technical Department included the Technical Divisions and Research Committee; the Administrative Department encompassed the Membership Qualifications, Local Sections, Student Chapters, and Junior Member committees; and the Professional Department comprised new committees dealing with engineering education, public relations, legislation, registration of engineers, fees, and salaries.

More significant is the Legislation Committee charged with recommending

> to the Board of Direction such action as is deemed advisable with respect to legislation contemplated or in process, an engineering

> analysis of which will be calculated to be of help in the determination of a solution beneficial to the public.

Finally, 78 years after its organization, ASCE defined a professional development policy, and it set up the machinery to carry it out. The educational work begun in 1907, the registration activity initiated in 1910, and the ethical responsibilities assumed in 1914 now had systematic direction. The means were in place to administer standards of professional education, professional competence, and professional practice, and to serve the public interest.

The timing was unfortunate. The Great Depression began in 1930, producing complex economic and manpower stresses in the United States, and in ASCE. Major emphasis was given to the economic welfare of the membership, and funds for new activities were limited. Nevertheless, the 1930s were the turning point in ASCE professional development. In 1934—the very depths of the Depression—a Committee on Aims and Activities was charged to study current programs "with respect to human relations, welfare, and public relations."

The creation of the National Society of Professional Engineers (NSPE) in 1934 was no doubt a stimulus to the emphasis given professional affairs in ASCE during this period. The AAE, which had been an influence in 1919, was declining as a significant factor in the profession.

A 1937 member petition called for the creation of a Professional Activities Division to deal "with subjects incident to professional practice and ethics, promoting understanding among engineers and between the profession and the public, and increasing the usefulness of the profession." Particular reference was made to "the humanities of the profession," apparently in recognition of social concerns.

The petition resulted in establishing the Committee on Professional Objectives, with broad powers to coordinate the work of the whole spectrum of professional committees. The new committee would "strengthen and amplify the activities of the existing committees of the Society, and not to weaken or hamper such committees in any respect."

The committee undertook its charge by staging a special program for the Annual Meeting in 1939. A panel of high-ranking spokesmen for the American Institute of Architects, the American Medical Association, the American Bar Association, and ASCE compared the progress being made to improve the social, economic, and professional status of their respective groups. Committee Chairman Enoch R. Needles (President, 1957) observed:

> For 87 years, we have functioned almost exclusively as a technical society. Whether we are architects, physicians, lawyers, engineers, educators, or clergymen, we must acknowledge certain common fundamentals. A high degree of education for the professional man is taken for granted. His actions must comply with high ethical standards. His

endeavors are largely devoted to service to his fellow man. He seeks respectability, and aspires to be deserving of respect. Finally, he desires a reasonably adequate compensation for his labors.

The Committee on Professional Objectives, which continued until January 1947, prompted greatly increased professional development. Action programs were especially effective in the domain of economic welfare of the engineer, covering civil engineering salaries, collective bargaining, unionization, and consulting fees.

The 1947 coordination effort reconstituted the Committee on Professional Objectives to include the Board Contact Members of the seven professional committees. Then, in 1953, this group of functions was designated as the Department of Conditions of Practice, to set it apart from the Department of Technical Activities. A 1955 study clarified Society policy on practice issues and reaffirmed the collection and dissemination of guideline salary data and the recommendation of salary schedules in a "professional way" (without resort to trade union tactics).

By 1958, the review, reappraisal, and reorganization of professional functions had become a continuous process. The programs were often impeded by bureaucratic detail. The Board then created a Task Committee on Administrative Procedure "to study and review the responsibilities and assignments of members of the Board ... with a view toward a reduction of the routine functions now imposed on officers and directors." Their 17 recommendations urged eliminating some committees, reducing the size of others, and simplifying procedures in general. Some recommendations were implemented, but the effect of this exercise was temporary, at best.

ASCE has always been exempt from income taxes under the section of the Internal Revenue Code that applies to religious, scientific, charitable, educational, and literary organizations. A further requirement under this section, however, is that "No substantial part of ... activities may be devoted to the carrying on of propaganda or otherwise attempting to influence legislation." Unlike the other Founder Societies, ASCE had never viewed this regulation as a constraint on its professional or public service activities, even though the Society had been reclassified for a short time in 1948 because of a misunderstanding with the Internal Revenue Service (IRS). The 1961 "Legal Audit," conducted by a Washington, D.C., law firm, analyzed all current Society programs with respect to IRS requirements.

The investigation concluded several policies and programs were subject to legal questions. After careful deliberation, the Board of Direction decided that the Society would continue to act as necessary to serve the public interest and meet its responsibilities as a first-order professional organization. Should there be IRS questions under the terms of Section 501(c)(3) of the Internal Revenue Code, a decision would then be made as to accepting the less favorable but less confining tax exemption status under Section 501(c)(6)as a "Business

League." A 1966 reorganization of Society committees replaced the Department of Conditions of Practice with a Department of Professional Activities, without significant change in committee functions.

In 1968, the ASCE Board appointed an ad hoc Professional Activities Study Committee—despite the existence of the standing Committee on Society Objectives, Planning, and Organization (COSOPO)—to evaluate the Society's professional objectives and to make recommendations to the Board of Direction. This task force triggered another reorganization in 1970, but it also brought about an increased commitment of Society resources to professional development activities. As reorganized, the Professional Activities Committee was delegated to administer four new Professional Divisions, each with its own executive committee and these subcommittees:

- The Administrative Division included a Budget Committee on Professional Activities and committees on Professional Prizes and Awards, Professional Publications, Programs for Professional Sections, and Public Relations.
- The Education Division had subcommittees on Continuing Education, Curricula and Education, Career Guidance, Educator/Practitioner Interchange, Student Chapters, and Technician and Technology Education.
- The Member Activities Division set up committees on Legislative Involvement, Local Sections, Minority Programs, Public Affairs, and Younger Members.
- The Professional Practice Division covered Employment Conditions, Employer–Engineer Relationships, Pensions, Salaries, Unionization, National Salary Guidelines, Engineering Management, Registration of Engineers, Standards of Practice, Contingent Fees, Turnkey Contracts, and Engagement of Professional Services.

This list represents great progress from the halting efforts of the early years, but the administrative process was complex. For example, a proposal concerning education policy would be formulated in the Curriculum and Education Committee, after which the proposal had to be reviewed by both the Education Division's Executive Committee and the Professional Activities Committee before it could be presented to the Board of Direction.

At the Annual Meeting in October 1969, a new aspect of professional concern arose when a young African-American member, Ralph E. Spencer, challenged the Society for its lack of involvement in social issues. He referred specifically to poverty and racial inequality as areas for action by ASCE.

The Board of Direction responded, acknowledging the Society's need to recognize social impact and humanitarian service in civil engineering practice. It created the Committee on Minority Programs in 1970 to encourage and assist racial minorities to find successful careers in civil engineering.

During this period, COSOPO was completing (1973) a three-year appraisal of "Goals of ASCE," as follows:

Goal 1—To Serve the Public: To provide a corps of civil engineers whose foremost dedication is that of rendering a public service.

Goal 2—To Advance the Profession: To improve the technical capability and professional dedication of civil engineers.

Goal 3—To Improve the Status of Civil Engineers: To continue to improve the professional stature and economic status of civil engineers.

Goal 4—To Improve ASCE Operations: To devote resources, organization, personnel, and operation in the efficient pursuit of these goals of serving the public, the profession, and the status of civil engineers.

The Society struggled for almost 80 years to define its professional-development policy and then devoted the next 40 years to devising the mechanism for implementing that policy. At the same time, ASCE was initiating and developing standards of professional education, competence, and practice—as well as serving the public interest, as befits a professional organization.

The "Learned Art"

"A profession is the pursuit of a learned art in the spirit of public service." This succinct definition by Dean Roscoe Pound became a prime guideline for the ASCE leadership after it was quoted in a meeting of the Metropolitan Section in the late 1950s.

The idea of engineering as a learned art was recognized in the United States as early as 1815, when Captain Sylvanus Thayer began his one-man crusade to develop a sound basic curriculum and able faculty for the U.S. Military Academy at West Point, New York. Because of the practical orientation of civil engineering in 1852, however, ASCE did not undertake the direction of civil engineering education. The 1870 requirement for admission to membership—equating graduation from a "school of recognized standing" with two years of practical engineering experience—was a grudging acknowledgment of the value of academic training.

The educational backgrounds of the first members of ASCE varied. Biographical data on 27 of the 55 members listed in the first annual report show that of 11 who attended West Point, nine graduated. Ten attended other colleges; six earned at least one degree (several earned more than one); and six had no college education. For those without a college degree, modest formal education was augmented by self-study, and technical education and training were acquired through apprenticeship on engineering projects of the day. Among them were James P. Kirkwood, James Laurie, and William J. McAlpine, all of whom later served as ASCE Presidents. John B. Jervis, Hon.M.ASCE, spoke about engineering education in an address at the first Annual Convention of ASCE in 1869. Jervis, then 74 years old, built a distinguished career on a common-school education and 15 years of practical experience ranging from axman to Superintending Engineer on construction of the Erie Canal.

"After a fair education in the ordinary elements," said Jervis, "an aspiring engineer should study mathematics ... to make any computation of quantities, and to carry forward any investigations that he may find it necessary to make in pursuing the science of mechanical philosophy." Study in mechanical philosophy, "in which special attention should be paid to the character of all the materials required in the various structures ... and the form and position of materials best adapted to the end it is sought to serve," would be followed by courses in hydraulics, surveying, and the study of various structures under experienced engineers. At this point, the student would enter on the field under professional direction.

But there was much more to Jervis's curriculum. He gave great emphasis to the point:

> No skill in forming lines and levels, and in devising structures, will complete the education of an engineer without an intelligent capacity for conducting business. This is an important item in his education, and indispensable to a successful practice.

There was considerable difference of opinion about such education, just as there was a hundred years later. A paper by Thomas C. Clarke, presented in June 1874, noted the differences between the European and American approaches:

> The methods of this education used in this country and in England differ materially from those on the continent of Europe. Here we begin by a course of study at some school, college, or technological institution, and complete our education by serving as assistants on some class of public works, for which we receive more or less payment. In England, the course is the same except that there boys begin younger and with less perfect theoretical education; then they serve their time under some practicing engineer from three to five years, paying premium of £300 to £500 for the privilege. This they do, because their chance of future employment depends on being personally known to some engineer in large practice. On the continent of Europe ... education begins at the other end, by the compulsory acquirement of a high degree of theoretical knowledge. Partly with this and partly afterwards, a certain amount of practical information is given, but the leading idea is to make a man a thorough engineer theoretically before he begins or is even allowed to practice.
>
> In this country ... we have attempted with a wise eclecticism, to combine the advantages of both systems, and educate our engineers in the theoretical principles of the science first and then let them acquire practical knowledge by practice itself.

Clarke favored a classical education, strong in both natural science and the humanities, with mathematics limited to the ordinary analysis. Study of ancient and modern languages was urged, to improve communication capability. After such study, Clarke recommended that the student go into the field and office to learn about actual practice before returning to the technological school for training in his chosen field of specialty. The paper elicited spirited discussion.

Some engineers believed that inserting an internship between the general and the technical parts of the academic experience was not feasible and recommended that specialization be undertaken after entrance into practice, but that technical courses should be taught "by practical men—those who have worked in the field or shop and know what is wanted there."

All commenting agreed on the need for practical experience in connection with academic training. The divergence of opinion concerning curriculum balance among mathematics, natural sciences, and the humanities still continues today. Similarly, the present division between engineering and technology continues.

Early Engineering Education

An early manifestation of interest in engineering education was a visit during the 1875 ASCE Convention, held in New York City, to the Stevens Institute in Hoboken, New Jersey. Stevens—founded only four years before, in 1871, to train mechanical engineers—was outstanding at the time for its faculty, facilities, curricula, and research programs.

In 1874, Professor Estavan A. Fuertes, M.ASCE, of Cornell University, urged the Society to declare what should be the course of instruction for students of engineering. The request was referred to a special committee comprising Professor DeVolson Wood of Stevens Institute, Professor George W. Plympton of Polytechnic Institute of Brooklyn, and Charles MacDonald. The committee's report, adopted on May 6, 1874, stated in part:

> We are of the opinion that the Society is not an advisory body in such matters. ... the institutions of learning should be left free to construct their courses, and no attempt should be made to mould them after a fixed pattern.
>
> We ought not to import a foreign system, but seek to build one here especially adapted to our times and circumstances. Our schools ought not to graduate men with a mere "smattering" of the sciences they are to use, but the instruction should be thorough and the standard of graduation high. If an error is made in either direction, the schools should be too theoretical rather than too practical.
>
> It makes but little difference what degree is conferred at graduation; but if that of "Bachelor of Engineering" instead of "Civil Engineering" on the ground that the latter implies a certain amount of practical experience, then we respectfully submit that the latter

> should be conferred only by a body of practical engineers; such for instance as this Society. In the sense, however, that the candidate has acquired a thorough knowledge of the "Science of Engineering," we see no impropriety in conferring the latter at graduation. The student can become a "Master of Engineering" only by long and varied experience.

Polarization of proponents of the "practical" and the "scholastic" elements of the profession was raised by the joint ASCE–AIME Committee on Technical Education, possibly the first intersociety activity in engineering education. A program arranged by the joint committee during the 1876 Centennial addressed two questions:

1. Should a course of practical instruction precede, accompany, or follow that in technical schools?
2. Is it practicable to organize practical schools under the direction and discipline of experts in engineering works?

Although these questions were not answered, the airing of conflicting views was salutary.

Acting on a July 1887 resolution, the Board approved an ad hoc Committee on Professional Training and Technical Education to review the proposal for a new committee on technical education. In its 1888 report, this committee recommended that no action be taken at that time to expend Society funds "in discussing the early education of engineers so long prior to the time when they can, under any circumstances, become members of the Society."

When the Board adopted this recommendation and so reported to the membership, Oberlin Smith, M.ASCE, disagreed, saying:

> Members here have no idea of the immense importance of the education business; we think these students don't amount to much. Fifteen or twenty years hence, when a good many of us are dead, these young men are going to be in the places that we are in now.

It was exactly 20 years later that the next significant action occurred in the field of engineering education.

Education and Professional Responsibility

At the 1893 Columbian Exposition in Chicago, ASCE and other engineering societies staged the Engineering Congress. The sessions conducted on engineering education generated so much interest that the Society for the Promotion of Engineering Education (SPEE) was organized almost immediately afterward. SPEE later became the American Society for Engineering Education (ASEE). Professor DeVolson Wood of Stevens Institute of Technology, the first president of SPEE, was a member of ASCE from 1872 to 1887. Other SPEE officers were also ASCE members.

A few months after a special Committee on Engineering Education was appointed in 1907, SPEE invited ASCE to participate in a joint committee with ASME, AIEE, AIME, and SPEE to study engineering education and report upon its "proper scope and proper direction for advance." After some reluctance, the Society designated two members to serve in the joint effort. In 1910, the Board granted the ASCE Committee on Engineering Education $200 to compile certain statistics gathered by the Carnegie Foundation. In the meantime, the Board of Direction rather brusquely declined a proposal from SPEE that the latter organization hold its annual convention at the same time and place as ASCE. This would seem to have been a fine opportunity for interaction and collaboration, as was the case 20 years later when ASCE contributed its cooperation, support, and $2,500 to a Summer School for Engineering Teachers sponsored by SPEE at Yale University.

In 1912 the Committee on Engineering Education reported upon a study "of the work of instruction as carried on by 20 leading technical schools and colleges in the United States." When the data were compiled, an evaluation meeting was held where attendees expressed different views of engineering education, and advocated and dissected reforms. The committee reported in part that

> the whole matter of Engineering Education will be taken up in a scientific manner by the Carnegie Foundation, which recently did the same thing for Medical Education, in which they expended $40,000, with grand results—rendering an inestimable benefit to that branch of scientific instruction.

Professor C.R. Mann directed the study, under the auspices of representatives of the four Founder Societies, along with SPEE and the American Chemical Society. The results appeared in 1919.

A surprising difference of opinion between educators and practitioners was revealed by the Carnegie appraisal. A questionnaire circulated to members of the engineering societies brought 1,500 replies, from which characteristics for engineering were weighted; see Table 4-1.

Although the validity of such data might be questioned, Mann drew an interesting conclusion from the survey:

> If this is the correct definition of an engineer, the schools must reorganize very fundamentally. Very little conscious attention is there being paid to those qualities on which 87% of the success of the engineer depends—character, judgment and efficiency. These personal qualities are usually regarded as a sort of by-product ... the development of these important personal qualities is not in the focus of attention of the school. Their attention is focused on the last two

Table 4-1. *Characteristics Needed to Become a Successful Engineer, According to Survey Results in 1919*

Characteristic	*Weight (percent)*
Character (integrity, responsibility, resourcefulness, initiative)	41
Judgment (common sense, scientific attitude, perspective of life)	17.5
Efficiency (thoroughness, accuracy, industry) and Understanding of Men (executive ability)	28.5
Knowledge of Engineering Science Fundamentals	7
Technique of Practice and of Business	6

> items, which have a weight of 13 per cent. If you engineers … will define the product for the schools, the schoolmen will prove competent to produce it.

The final report dealt with the prevailing status of education in 20 schools, each of which Mann visited. Among the problems he found were that only 40% of all engineering students completed their course, that "specialization and subdivision of curricula has gone too far," and that there was wide inconsistency in course content, testing, and grading, and in the utilization of shop work. His recommendations included the conduct of research on aptitude tests and student performance, reduction in the number of courses to four or five each term, and recognition of professional experience as a qualification for teaching. Noting that "the spirit of research is part of university life," he urged that teaching and research be conducted in such balance as to realize the best values of both. The inculcation of professional attitude and discipline in the student was held to be a responsibility of the school and industry, and experimentation was needed on how this might best be done.

The Board of Direction accepted the report, and discharged the Committee on Engineering Education after its 12 years of existence.

For the next several years, although ASCE did not undertake independent educational activity, it was closely associated with SPEE. Under another Carnegie Foundation grant to SPEE, a significant "Report of the Investigation of Engineering Education (1923–29)" was produced with ASCE's support.

As the number of engineering schools increased, questions began to arise about the interpretation of the term "recognized standing" as it was used in the membership requirements of the Society. Since 1928, graduation from a "school of recognized standing" had been equated with four years of active practice. In 1930, the Board of Direction appointed a Committee on Accredited Schools to maintain its own list of accredited civil engineering schools; this was done until 1938.

A giant step forward was taken by the engineering profession on October 3, 1932, when the Engineers' Council for Professional Development (ECPD) was formed to accredit engineering schools and enhance the professional recognition of engineers. Included in the original membership were ASCE, AIME, ASME, AIEE, SPEE, the American Institute of Chemical Engineers (AIChE), and the National Council of State Boards of Engineering Examiners (NCSBEE).

In 1938, the ASCE Committee on Accredited Schools recommended that the Society adopt the accrediting service of ECPD, with suitable provision for the schools formerly accepted by ASCE but not by ECPD. The ASCE committee was then discharged.

In 1965, ASCE produced the "Guide for Civil Engineering Visitors on ECPD Accreditation Teams," which then served as a model for other ECPD societies.

World War II brought unusual problems to the colleges and universities of the United States. In January 1943, the Board of Direction adopted two significant resolutions. The first of these, "Emergency Conditions in Engineering Education," urged that the federal military agencies and War Manpower Commission maintain a strong basic engineering education and student body to ensure successful prosecution of the war as well as postwar reconstruction and rehabilitation. Attempts to convert liberal arts and nonscientific curricula into "so-called engineering courses" were opposed.

The second resolution, "Long-Range Problems and Objectives of Engineering Education," affirmed the conviction

> that active participation in defining and promoting the basic educational standards of the profession of civil engineering is a necessary and proper function of the Society, and that such participation is hereby adopted as policy of the Society.

This position is almost diametrically opposite the attitude of the leadership of the Society in 1874!

The resolution went on to authorize a joint study of civil engineering education with the Civil Engineering Division of SPEE. The first part of the study, reported in 1944, was based on a detailed study of civil engineering curricula in 114 schools.

The curriculum in Table 4-2 recommended for the four-year bachelor's degree program is a milestone in that it incorporates parallel development of the scientific–technological and the humanistic aspects of engineering education.

Although the 1944 report was under review by engineering school administrators, ASCE conducted a survey of 2,700 members regarding civil engineering education needs. The results appeared in a second report in 1945. There was a 38% response to the survey, which judged graduates of the 1935–45 decade to be lacking in oral and written communication skills, and to have little interest in public affairs. Those engineers were rated favorably in other areas—the ability to get along with others, leadership capacity, grasp of fundamentals, and qualities of accuracy, diligence, and dependability. The survey endorsed the curriculum structure recommended in the 1944 report.

Table 4-2. *Recommended Curriculum for Bachelor's Degree Program in Engineering in 1944*

Subject	*Percentage of Total Coursework*
Humanities/social	20
Physical sciences, including geology	15
Drawing	4
Mathematics, not including trigonometry	10
Mechanics, hydraulics, strength of materials	11
Engineering subjects other than civil	10
Civil engineering	30

Both reports were published (*Proceedings*, March 1946) and widely circulated. Because many engineering schools had low enrollments at the close of World War II, they were in a good position to adjust their programs.

An Exploratory Conference on Engineering Education in 1947 brought together the chairs of all ASCE committees related to the subject, together with representatives from ECPD, ASEE, and the staff. Recommendations from the conference urged that ECPD expand its guidance program and implement the 1945 ASCE curriculum recommendations; that Local Sections promote part-time employment opportunities for engineering teachers; that salary studies and guidelines be provided for engineering educators; and that ASCE professional activities be coordinated toward these goals. The Committee on Salaries promptly carried out the survey of qualifications, responsibilities, and compensation of civil engineering teachers, reporting to the Board of Direction in 1948.

The Committee on Engineering Education found in 1948 that many young graduates with advanced degrees were becoming teachers without having acquired practical experience. The committee asked the Local Sections to find employment for young teachers and to encourage university administrators to release them for professional summer work.

New Trends in Education

In 1953, recognizing the need to study civil engineering education independently, Dean F.M. Dawson of the University of Iowa, convinced the Board of Direction that a Task Committee on Civil Engineering Curricula should be formed. In 1954, the Board authorized the Task Committee on Professional Education. The Board approved an appropriation of $15,000 for the committee to conduct an opinion survey with the condition that the survey would cover areas other than education. A Task Committee on Economic Advancement Objectives cooperated in the survey.

The task committee report, including the survey results, was published in *Civil Engineering* (February 1958). Twelve very broad recommendations were set forth, seven of which pertained to public relations, economic welfare, unionism, and Local Section activity. The education-related findings urged higher salaries for teachers, attention to the needs of the construction industry, more emphasis on career guidance, and stronger postbaccalaureate programs. Most important, however, was the following recommendation:

> *Recommendation:* Although certain undergraduate courses are commonly required for all students in engineering, civil engineering curricula must maintain their separate identity and objectives and the civil engineering departments should receive support consistent with the civil engineer's great contribution to society.

While the Task Committee on Professional Education was at work, a general study was completed by ASEE under the direction of Professor L.E. Grinter, Hon.M.ASCE. This "Evaluation of Engineering Education (1952–55)" had a profound effect on the sophistication of all engineering education by its encouragement of stronger training in mathematics and the sciences, and in the application of scientific fundamentals to engineering problems. This was in part a natural result of the postwar acceleration of technological progress.

Through the initiative of Professor Felix A. Wallace, F.ASCE, of the Cooper Union, New York City, a major Conference on Civil Engineering Education was held at Ann Arbor, Michigan, July 6–8, 1960. The Cooper Union, ASCE, and ASEE jointly sponsored the event, with funding assistance from the National Science Foundation. About 250 delegates and participants attended and at the conclusion, placed 11 resolutions before the conference. When the oratory and parliamentary jousting ended, the attendees agreed that nine of the 11 resolutions would be published; the thrust of these was:

1. That the Conference favors the growth of a preengineering, undergraduate, degree-eligible program for all engineers, emphasizing the humanities, social studies, mathematics, basic and engineering sciences, with at least three-quarters of the program interchangeable among the various engineering curricula; to be followed by a professional or graduate civil engineering curriculum based on the preengineering program and leading to the first engineering degree, with a civil engineering degree awarded only at the completion of the professional or graduate curriculum. Further, that increasing opportunities be provided for qualified students to earn graduate degrees at the master's and doctor's levels.
2. That the four-year undergraduate program with the B.S. degree in specific fields of engineering be retained.
3. That the "Civil Engineer" degree as a nonresident, nonacademic degree be abandoned.

4. That the Conference favors establishing graduate professional schools offering the degrees of Master of Engineering and Doctor of Engineering in several engineering specialties, and favors continuing programs leading to the Master of Science and Doctor of Philosophy degrees.
5. That the Conference favors development in Colleges of Arts and Sciences of a three-year undergraduate preengineering program for all engineers, emphasizing the humanities, social studies, mathematics, basic science and communications, followed by a three-year engineering program of engineering science and professional courses taught in professional schools of engineering, extending about as far as present Master Degree programs and leading to a professional degree in engineering.
6. That it is the responsibility of the practicing engineers to achieve professional standards of performance and ethics.
7. That the civil engineering educator take a position in the profession to maintain an appreciation among engineers and the public of the significance of basic scientific and general cultural education.
8. That the Conference favors the growth of graduate programs in schools; that each institution respond to the needs of students, parents and employers … and that experimentation be encouraged and special programs perfected for gifted students.
9. That the time has come to increase the length of the curriculum for the first degree in Civil Engineering from four years to five years.

These far-reaching proposals were incorporated into specific guidelines in 1962 by the Committee on Engineering Education, following a survey of 112 ECPD-accredited civil engineering programs. Among the policy recommendations were a five-year first degree combining a preengineering core with a professional program and at least two types of postgraduate programs—one for those pursuing careers in academia or research and development, and the other serving those wishing to be practitioners in a specialty field.

These guidelines lasted for four years, and they were applied at least in part in a number of schools. In 1966, a major appraisal of all engineering education was made by ASEE with National Science Foundation funding of $307,000. This study, "Goals of Engineering Education," involved 4,000 questionnaires sent to engineering practitioners and others.

One study proposal drew opposition from ASCE's Board of Direction: "The first professional degree in engineering should be the Master's Degree, awarded upon completion of an integrated program of at least five years' duration." Such degree would not identify any field of specialization, and the Board disagreed strongly with the idea of shaping all engineers by the same template. In 1968, ASCE replaced its 1962 policy with one highlighting the need to educate the civil engineer to be able to function in cooperation with other design professionals to solve complex environmental problems. The policy also called for four years of study leading to a bachelor's degree from an accredited civil engineering to be the minimal education for the civil engineering profession. A fifth year was minimal for civil engineering specialists.

Costly as it was, the Goals Study failed to recognize an emerging development that rendered its findings obsolete almost as soon as they were published. The post-Sputnik years had created a shortage of engineers capable of handling practical planning, design, and construction operations. As a result, several schools introduced four-year bachelor of engineering technology (BET) programs. By the mid-1970s, there were several thousand BET graduates annually. The registration process somehow had to accommodate the BET.

In 1972 the ASCE Committee on Engineering Education was replaced by an Education Division, with six committees: Continuing Education, Curricula and Accreditation, Guidance, Interchange between Education and Practice, Student Chapters, and Technician and Technology Education. The new Division undertook another appraisal of civil engineering education by sponsoring the ASCE Conference on Civil Engineering Education at Ohio State University in February 1974. The conference attendees focused on current developments as the BET, the environmental assessment of engineering functions, and the growing demand for multidisciplinary interaction.

Though ASCE sponsored the conference, 17 other civil engineering–related organizations participated. Of the 350 conferees, a third were noneducators—an unprecedented ratio for such gatherings.

Continuing Education

In the 1960s, ASCE's education concerns were expanded beyond policy and curriculum development. The Society began in 1961 to sponsor regular regional meetings of the heads of civil engineering departments. These were immediately successful, and were followed in 1966 by the scheduling of breakfast gatherings of department heads and ECPD accreditation team visitors at all national ASCE meetings. Committee and staff participated in these discussions.

In 1966, an Assistant Secretary for Education joined the staff. Society policy for continuing education favored collaborating with existing programs in schools, engineering societies, and industry rather than creating competing programs. Emphasis was on seminars in the Local Sections and at national meetings of the Society, usually under joint sponsorship.

ASCE also offered self-study courses, beginning in 1970. The first of these was "Engineering Economy." The Society also encouraged self-study by providing bibliographies of civil engineering books, on both technical and professional subjects.

In 1970, ASCE compiled guidelines to assist Local Sections in continuing education. One success was the seminar assembled by the American Institute of Steel Construction on "Plastic Design of Braced Multi-Story Frames," presented through ASCE Sections in 50 cities. A list of current seminars, conferences, and short courses appeared regularly in *Civil Engineering* magazine, and a professional continuing education specialist joined the staff in 1973.

Career guidance services were modest prior to 1958.

In the late 1950s, static civil engineering enrollments inspired an ASCE career brochure, which encouraged junior high school students to take the courses necessary for engineering. ASCE spent about $24,000 to produce and distribute 500,000 copies, and it also published a paperback book, *Your Future in Civil Engineering*, by Alfred R. Golze, F.ASCE. It cost the Society $30,000 to produce 200,000 copies of this book, which were distributed free of charge. In 1967, a colorful illustrated brochure, "Is Civil Engineering for You?," became ASCE's primary career guidance literature. About 250, 000 copies were distributed at a cost of $50,000; a new edition was produced in 1974.

The 1969 color film, *A Certain Tuesday*, was shown in high schools, with narration preferably by a civil engineering student who was an alumnus. The film was widely used by Local Sections. In 1971 ASCE took part in a guidance program called "The World of Construction." This junior high school–level industrial arts course aimed at involving students in construction activities to arouse career interest. More than 1,500 schools offered the course in 1973.

A 1965 Task Committee on Civil Engineering Management Education recommended that ASCE provide partial funding for pilot courses in civil engineering management in selected universities. Fellowships of $2,000 per year for five years were set up at several universities—Carnegie–Mellon, Drexel, Illinois, Purdue, and Stanford. The program ended in 1972.

This heavy financial commitment to education was possible only because of the "Voluntary Fund," contributions and annual dues from Life Members who are no longer obliged to pay. The fund began in 1947; by 1974, it had made about $300,000 available for the special needs of the Society.

The educational efforts became international in 1958, when ASCE assumed responsibility for the secretariat of the Committee on Engineering Education and Training of the Conference of Engineering Societies of Western Europe and the U.S.A., known as EUSEC. The Executive Secretary of ASCE at the time served as General Secretary of EUSEC as well as secretary of its education committee. ASCE provided fund-raising and administrative services and offered its professional experience.

About $60,000 was raised to produce a classic comparative study of engineering education in 19 countries. This three-volume *Report on the Education and Training of Professional Engineers* (1960), published in English, French, and Spanish, provided detailed information about the general education systems in each country, the university level systems, practical training requirements, graduate education, and criteria for professional recognition. The report, funded by the Ford Foundation, the Organization for European Economic Cooperation, and the various EUSEC societies, has provided guidance to many international studies on engineering education since its publication in 1961.

Education Evolution during 123 Years

The changes in engineering education since ASCE was organized are evident. The 1852 and 1973 courses of study are compared in Figures 4-1 and 4-2. The 1973 Professional Program at Rensselaer requires five years for completion—the same three-year Pre-Engineering Curriculum plus two years of civil engineering and electives in the humanities leading to the master of engineering degree. The Pre-Engineering Curriculum is common for all branches of engineering, and the engineering electives are mainly in the areas of engineering science, economics, basic science and management. Conversely, the 1852 Course of Studies was rigorous and surprising in its breadth and in its coverage of mathematics and the sciences.

While ASCE was conservative in many policy areas, such was not the case with education. As early as 1874, the Society encouraged innovation, and it studied trends in civil engineering education a dozen times between then and 1974.

Since 1940, there has been a growing awareness that civil engineering education must include social and political dimensions. After 1960, the need for civil engineers to be able to assess the impact of their works on the environment added another aspect to their education. The kind of engineer sought by "industry" was not necessarily prepared to plan, design, build, and manage public works to satisfy 1974 standards. Only through periodic self-appraisal can a profession maintain an educational system that will advance its "learned art."

Membership Requirements

Because the Institution of Civil Engineers of Great Britain (ICE) was so influential in the organization of ASCE, membership standards of the Society limited the higher grades to those having education and experience identifying them as professional-level engineers. ICE was legally chartered to serve this function; ICE membership automatically provides a legal right to practice under the terms of the organization's Royal Charter.

ASCE was not so ordained, but the Society's early leaders intended membership to be proof that an individual is fully qualified to practice civil engineering. "Professional identification," a tangible value to the early member, was recognized by the courts as well as the public.

When legal registration was first proposed in 1897, many members were opposed. They felt that ASCE membership was sufficient to set the qualified civil engineer apart from the incompetent practitioner. Support for engineering registration came from Local Sections of the Society.

Although membership admission procedures were quite informal in the beginning, there was still a high order of selectivity exercised. The initial "grandfather" privilege was extended to an invited list of engineers of national reputation in the following terms:

Division	Departments of Instruction.	Subjects of Study and Practical Exercises.
FIRST YEAR.		
DIVISION C—THIRD CLASS.	MATHEMATICS.	Algebra.—Elementary Geometry.—Nature and use of Logarithmic and Trigonometric Tables.—Trigonometry.—Mensuration.
	PHYSICS.	General Properties of MATTER.—Nature of the Physical FORCES.—Phenomena and Laws of GRAVITY.—Phenomena and Laws of HEAT.
	GRAPHICS.	Use of Drawing Instruments.—Graphical Constructions of Chain and Compass Surveys.—Copying of Mechanical Drawings.
	CHEMISTRY.	Principles of Chemical Philosophy.—Study of the Non-Metallic Elements.—Laboratory Practice.
	GEODETICAL ENGINEERING.	Operations in the Field,—Measuring of Lines; Chain and Compass Surveys of Fields and Farming Estates; Dividing Land.—Computations of Areas, etc.—Mapping of Surveys.
	PHYSICAL GEOGRAPHY.	Structural and Systematic Botany.—Vegetable Physiology.—Geographical Distribution of Plants.
	ENGLISH COMPOSITION AND CRITICISM.	Section Lecture Exercises with Criticisms.—Keeping of Lecture Books.—Writing of Special Theses, etc.
	FRENCH LANGUAGE.	Elements of French Grammar.—French Exercises and Translations.
SECOND YEAR.		
DIVISION B—SECOND CLASS.	MATHEMATICS.	Analytical Trigonometry.—Analytical Geometry.—Differential Calculus.—Integral Calculus.
	PHYSICS.	Terrestrial Magnetism.—Statical and Dynamical Electricity.—Electro-Magnetism.—Magneto-Electricity.—Acoustics.—Optics.
	GRAPHICS.	Descriptive Geometry.—Measuring and Sketching Engineering and Architectural Structures, and Construction of Working Drawings from these data.—Topographical Drawing.
	CHEMISTRY.	Chemical Study of the Metals.—Exercises in Qualitative Analyses.
	GEODETICAL ENGINEERING.	Theory and Adjustments of Field Instruments.—Trigonometrical Determination of Heights and Distances.—Field Exercises in General Geodetic Operations.—Field Practice in Hydrographical Surveying.
	GEOLOGY.	Descriptive Mineralogy.—Systematic and Descriptive Geology.—Economic Geology.
	ENGLISH COMPOSITION AND CRITICISM.	Section Lecture Exercises with Criticisms.—Writing out of General Lectures.—Writing of Theses on Scientific and Practical Subjects.
	FRENCH LANGUAGE.	Double Translations, (French and English.) Reading of French Scientific Authors.
THIRD YEAR.		
DIVISION A—FIRST CLASS.	RATIONAL MECHANICS.	General Statics and Dynamics of Solids, Liquids, and Gases.
	PRACTICAL ASTRONOMY.	Investigation of Astronomical Principles and Data for the solution of the Practical Problems of the *Meridian*, *Time*, *Latitude*, and *Longitude* of a place.—Sextant and Transit Observations, including Computations for, and Reductions, made by students.
	CONSTRUCTIVE ENGINEERING.	Equilibrium and Stability of Architectural and Engineering Structures.—Materials for Construction.—Theory of Machines.—Road Engineering.—Hydraulic Engineering.—Steam Engine and Hydraulic Motors.
	GEODETICAL ENGINEERING.	Higher Geodetic Surveying of Extensive Areas by Trigonometrical and Astronomical Methods.—Topographical Surveying, with Field Practice, Reductions, and Construction of Maps.—Surveys, Location, etc., of Engineering Works.
	PRACTICAL CHEMISTRY AND PHYSICS.	Chemical Study of the Principal *Economic Minerals*:—Blow-pipe Examinations.—Practical Exercises in Determining the Specific Gravties of Solids, Liquids, and Gases.
	PHYSICAL GEOGRAPHY.	Meteorology, General Hydrology, and Topography of the Earth's Surface.—Distribution of Plants and Animals.—Relations of Physical Geography to Engineering Works of Inter-Communication.
	GRAPHICS.	Descriptive Geometry, embracing Projections of Shades and Shadows, and the Principles and Practice of Natural and Isometrical Perspective.—Sketching of Machines, and Construction of Working Drawings of same.—Topographical Drawing.—Drawing Maps and Sections of Surveys for Lines of Transit.
	ENGLISH COMPOSITION AND CRITICISM.	SECTION LECTURES,—*Extemporaneous* and *Written*,—with full Criticisms.—*Written Reviews* of Machines, Structures, and Processes in the vicinity of Troy.—*Scientific* and *Practical Theses*.

Figure 4-1. *Rensselaer Polytechnic Institute: Course of Studies and Engineering Exercise for the Degree of Civil Engineer, 1952*

First Semester		*Second Semester*	
First Year			
Mathematics I	4	Mathematics II	4
Chemistry I	4	Physics II	4
Physics I	4	Humanities or social science elective	3
Elementary Engineering	3	Engineering electives (2)	6
Humanities or social science elective	3	Physical education or ROTC	—
Physical education or ROTC	—		
Second Year			
Mathematics III	3	Differential Equations	3
Physics III	4	Engineering I	3
Materials	3	Parameter Systems	3
Mechanics	4	Engineering elective	3
Humanities or social science elective	3	Humanities or social science elective	3
Physical education or ROTC	—	Physical education or ROTC	—
		Engineering seminar	—
Third Year			
Engineering Lab I	2	Engineering Lab II	2
Thermodynamics I	3	Humanities or social science elective	3
Mechanics	3–4	Engineering electives (4)	12
Humanities or social science elective	3		
Engineering electives (2)	6		
Engineering seminar	—		
Fourth Year			
One course each in structures, transportation, and soil mechanics and foundation engineering			9
One course from either construction, professional practice, environmental engineering, or water resources			3
Four additional civil engineering electives			12
Humanities or social science electives (2)			6

Figure 4-2. ***Rensselaer Polytechnic Institute: Baccalaureate Civil Engineering Curriculum, 1973***

> Such gentlemen only as receive this circular are eligible as members of the Society by a bare notice of their desire to become such and a compliance with the accompanying forms (on or before December 1st, 1852). All others will be elected by ballot in conformity with the requirements of the Constitution.

These requirements were not definitive: "Civil, geological, mining and mechanical engineers, architects, and other persons who, by profession, are interested in the advancement of science, shall be eligible as members." Processing of membership applications apparently involved only review and acceptance by those attending a meeting. At the first Annual Meeting, on

October 10, 1853, there were 55 members: 6 Honorary, 1 Corresponding, and 48 in the grade of Member. Most, if not all, of these are believed to have been on the original invitation list.

Membership actually declined by one in the year 1853–54, and the Society reduced dues of Resident Members from $10 to $5 and for Non-Residents from $5 to $3. At that time, only 10 of the 47 in the Member grade were Residents. The Board action was taken in the hope of attracting new members.

When the Society resumed activities in October 1867 after the Civil War, there was a healthy membership growth, despite the dues increase back to the pre-1853 rates. Fifty-four candidates were elected at the December 4, 1867, meeting, of whom 32 qualified by payment of dues. By 1870, total membership approached 200. The Society's membership became national by 1873, when about 70% of the total was Non-Residents (those living beyond 50 miles of New York City). Membership requirements are summarized in Table 4-3, which compares the qualifications for various grades at eight stages in their 122-year metamorphosis.

The increasing rigor of requirements is substantial, with the greatest changes between 1870–91 and 1950–72. The frequency of bridge and dam failures as a result of faulty design may have induced the emphasis on professional competency. Since 1950, engineering practice has reflected postwar sophistication in the membership requirements.

The "General Eligibility Requirements" essentially stayed the same over the 120 years. The most important change was in 1930, when the "equivalency of the engineering degree" requirement increased from two to four years of professional practice.

There was little change in the "Eminence Grade" status of Honorary Member, but nomination and election procedures were substantially modified over the years. The sanctity of this grade as the Society's "highest honor" was carefully preserved. A maximum of 20 Honorary Members was established in 1868; in 1972 there was no prescribed maximum, but no more than one Honorary Member for each 7,500 members was authorized to be elected in any one year. Only 259 Honorary Members were named between 1852 and 1974, an average of but two per year.

The "Advanced Professional Grade" reflects the progressive sophistication of standards of competence. The recognition since 1959 given to legal registration is very important. Only registered engineers were made eligible for the new Fellow grade in 1959, and the Society was the first of the Founders to recognize this distinction. ASCE's Board rarely exempted a candidate from this requirement unless he was a nonresident of the United States.

The "Professional Grade" of Associate Member/Member reflects increasing emphasis on education and legal registration, as does the "Entrance Grade" of Junior/Junior Member/Associate Member. The "Non-Engineer Professional Grade" or Associate/Affiliate was remarkably consistent until 1972. At that time, amendments eliminated the minimum age limits and automatically transferred to the Affiliate grade Associate Members who failed to advance to a professional grade in the allotted time. In 1972 the Affiliate grade for the first time accepted both engineers and nonengineers.

The Fellow grade, which existed from 1870 to 1950, was available to anyone who contributed to the Fellowship Fund, which was originally established to finance the publication of papers read before the Society. The fund was abolished in 1891, after which Fellowship fees were credited to the "permanent funds of the Society." The one-time Fellowship fee began at $100 and increased to $250 in 1873. The number of Fellows in this grade ranged from 43 in 1870 to 40 in 1900 to only 1 when the classification was abolished in 1950.

The Fellow grade established in 1959 was decidedly different, in that it represented a truly advanced professional engineering category of members. However, the grade was not conferred but rather attained by application. The grade of Corresponding Member was available from 1852 to 1877 to eminent foreign engineers who were willing to communicate with the Society at least once a year. Few Corresponding Members were enrolled, however, and after 1877 qualified foreign engineers were admitted as Non-Resident Members.

The Society maintained an "open" membership classification for those unable to qualify for other grades of membership for only four years, from 1891 to 1895. This "Subscriber" classification also found little demand.

ASCE has always considered the processing of membership applications a serious matter; the Society may well be the most "exclusive" organization to which many of its members belong. Since 1885, the candidate has been required to furnish a statement of his qualifications and to name endorsers who can verify his good character and reputation as well as his education, technical training, and professional experience. An official application form, requiring signature by the applicant and certified endorsement by his proposers, was adopted in 1872. Lists of candidates were first posted in advance of election; later they were mailed to all members, and from 1930 until mid-1968 they were published in *Civil Engineering*.

Originally, all membership admissions and transfers were voted upon by the entire body of members. In 1879, the election of Honorary Members was delegated to the Board of Direction; in 1890, the Board was authorized to elect applicants to all grades other than the Corporate (voting) grades of Member and Associate Member; and in 1909, the Board took over all membership functions.

By 1930, the volume of applications was such that the Board created the Committee on Membership Qualifications. In 1931, provision was made for Local Membership Committees to furnish "reports with respect to the admission or transfer of all applicants residing in the territory assigned to each local committee." These committees, renamed "Local Qualifications Committees" in 1946, have interviewed and reported upon thousands of membership candidates. They are believed to be unique in their function, at least in the engineering profession.

Table 4-3. *Evolution of ASCE Membership Requirements*

	General Eligibility Requirements	*Eminence Grade*	*Advanced Professional Grade*	*Professional Grade*
1852	Civil, geological, mining, and mechanical engineers; architects; and other professionals interested in advancement of science.	*Honorary Member:* Unanimous selection.	*Member:* General eligibility requirements above.	None.
1870	Same as 1852.	Same as 1852.	*Member:* Five years of engineering practice, including responsible charge; college diploma equivalent to 2 years of practice.	*Associate:* General eligibility requirements above.
1874	Civil, military, geological, mining, and mechanical engineers; architects; and other professionals interested in advancement of science.	*Honorary Member:* Eminence in engineering with 30 years of practice.	*Member:* Civil, military, mining, or mechanical engineer with 7 years of practice, or 5 years with C.E. degree, including 1 year in responsible charge.	None.
1891	None.	*Honorary Member:* Eminence in engineering or related science.	*Member:* Civil, military, naval, mining, mechanical, or electrical engineer or architect 30 years of age; 10 years of practice, 5 years of responsible charge; qualified to design; engineering degree equivalent to 2 years of practice.	*Associate Member:* Professional engineer or architect 25 years of age; 6 years of practice, 1 year in responsible charge; engineering degree equivalent to 2 years of practice.

1930	*Corporate Member:* civil, military, naval, mining, mechanical, electrical, or other professional engineer, architect, or marine architect. Engineering degree equivalent to 4 years of practice.	*Honorary Member:* Same as 1891.	*Member:* Age 35, with 12 years of practice, 5 years in responsible charge; qualified to direct, conceive, and design engineering works.	*Associate Member:* Age 27, with 8 years of practice, 1 year in responsible charge.
1950	Civil engineer or person qualified in engineering or allied profession. Engineering degree equivalent to 4 years of practice.	*Honorary Member:* Same as 1891.	*Member:* Same as 1930.	*Associate Member:* Age 25, with 8 years of practice, 1 year in responsible charge.
1959	Same as 1950.	*Honorary Member:* Same as 1891.	*Fellow:* Transfer from Member only. Age 40; legally registered engineer with 5 years in responsible charge as Member in engineering design or management.	*Member:* Age 27; 12 years of practice or engineering degree with 10 years of practice; 3 years in responsible charge.
1972	Same as 1950.	*Honorary Member:* Eminence in engineering or related art or science.	*Fellow:* Transfer from Member only. Legal registration as engineer or land surveyor, with 10 years in responsible charge as Member in engineering design, surveying, or management.	*Member:* ECPD-accredited degree, or approved degree with legal registration, plus 5 years of practice in responsible charge.

continued on next page

Table 4-3. **Continued**

	Entrance Grade	*Nonengineer Professional Grade*	*Financial Contributor*	*Foreign*	*Unrestricted Grade*
1852	None.	None.	*Fellow:* Contributor of funds to the Society.	*Corresponding Member:* Non-resident of the United States.	None.
1870	None.	None.	*Fellow:* Same as 1852	*Corresponding Member:* Same as 1852.	None.
1874	*Junior:* Two years of engineering practice or college degree and 1 year of practice.	*Associate:* Public works manager, scientist, industrialist, or other nonengineer professional qualified to cooperate with engineers.	*Fellow:* Same as 1852.	*Corresponding Member:* Same as 1852.	None.
1891	*Junior:* Age 18, with 2 years of practice or with an engineering degree. Must transfer to a higher grade by age 30.	*Associate:* Nonengineer professional qualified to cooperate with engineers in advancement of professional knowledge.	*Fellow:* Same as 1852.	None.	*Subscribers:* Persons age 21 not eligible to other grades (until 1895).

1930	*Junior Member:* Age 20, with 4 years of practice or with an engineering degree. Must transfer to a higher grade by age 33	*Affiliate:* Nonengineer professional; age 35, qualified to cooperate with engineers; 12 years of practice, 5 years in responsible charge	*Fellow:* Same as 1852.	None.	None.
1950	*Junior Member:* Age 20 with engineering degree or with 4 years of practice. Must transfer to a higher grade by age 32.	*Affiliate:* Same as 1930.	None.	None.	None.
1959	*Associate Member:* Age 20, with engineering degree or registration as Engineer-in-Training. Must transfer to higher grade within 12 years.	*Affiliate:* Engineer or nonengineer professional qualified to cooperate with engineers; age 27; with degree and 10 years of practice or with 12 years of practice, 3 years in responsible charge.	None.	None.	None.
1972	*Associate Member:* ECPD-accredited degree or approved degree with registration as Engineer-in-Training or master's degree. Must transfer to higher grade in 12 years or become Affiliate.	*Affiliate:* Graduate architect, scientist, lawyer, or other licensed or certified professional, or graduate in engineering technology with 5 years of acceptable experience.	None.	None.	None.

Note: ECPD is Engineers' Council for Professional Development.

Since 1950, staff verified the educational data in membership applications before referring them to the Committee on Application Classification. This committee—meeting for a full day each month—cleared those applications and submitted them to the Board for election by mail ballot after the references responded favorably. Where there were questions, the Board Committee referred cases to the Local Qualifications Committee for investigation and interview. Cases in doubt were then considered by the Committee on Membership Qualifications for specific recommendations to the Board.

This general procedure resulted in a declination rate of 6% to 10%. Most candidates failed due to deficient qualifications, but some applications were declined where the candidate's professional attitude or reputation was deemed not in accord with the objectives of the Society.

Table 4-4 shows annual dues in ASCE between 1852 and 1974. The difference in annual dues between Residents and Non-Residents prevailed until 1954, more than a century. Fees and dues were determined by membership referenda until 1970, when the Board of Direction assumed this authority.

An entrance fee has always been a requirement for ASCE membership. Since 1959, the $10 entrance fee has been waived for Student Chapter members who apply for Associate Member grade in the Society within a few months of their graduation. This inducement replaced the waiver of the first year's dues for "graduates of a school of recognized standing," in force from 1926 to 1959. During the Depression, ASCE waived the dues of some 2,033 members, beginning in 1932.

Since the Society's beginning, the Advanced Professional Grade of membership has held the Corporate privileges of voting and of holding office. ASCE also offered the Regular Professional Grade carrying these same privileges in 1891. Honorary Members not elected from the Corporate Member grades were given full privileges in 1950. The Entrance Grade had the right to vote in 1947, and the right to hold office in 1950. In 1950, the nonengineer Associate/Affiliate Grade received all membership privileges, but in 1959, the Society withdrew those.

An unusual feature of the ASCE membership requirements since 1891 has been the provision that a member in the entrance grade must transfer to a higher professional grade by a certain age or within a specified time. Several hundred members have been dropped each year for this reason.

A 1957 study showed the loss of entrance grade members was largely offset by lower annual loss rates in the higher grades of membership than that in other engineering societies. ASCE's membership grew from 55 members in 1852, to 2,221 in 1900, to 11,275 in 1925. In 1950, the Society's membership was 28,105; in 1974, it was 69,671. At this writing, only the Institute of Electrical and Electronic Engineers (IEEE) exceeds ASCE in membership among all the engineering organizations in the United States serving a single major engineering field.

ASCE has become one of the largest engineering societies in the world, despite its precarious beginning and its high admission standards. These

Table 4-4. *ASCE Annual Membership Dues (dollars)*

Membership grade	*1852*	*1853*	*1867*	*1868*	*1870*	*1872*	*1874*	*1891*	*1921*	*1930*	*1931*	*1950*	*1954*	*1971*	*1973*
M/F (R)	10	10	10	10	20	20	25	25	25	25	25	25	25	35	45
(NR)	5	3	5	5	10	10	15	15	20	20	20	20	25	35	45
AM/M (R)	—	—	—	—	15	15	—	25	25	25	25	25	25	35	45
(NR)	—	—	—	—	10	10	—	15	20	20	20	20	25	35	35[a]
J/AM (R)	—	—	—	—	—	—	15	15	15	15	15	15	15	20	25
(NR)	—	—	—	—	—	—	10	10	10	10	10	10	15	20	25
Assoc./Aff. (R)	—	—	—	—	—	—	15	15	15	20	25	25	25	35	45
(NR)	—	—	—	—	—	—	10	10	10	15	20	20	25	35	35[a]
Fellowship	—	—	—	100[b]	100[b]	250[b]	150[b]	250[b]	250[b]	250[b]	250[b]	[c]	—	—	—

Note: (R) denotes Resident; (NR) denotes Non-Resident Members. Grade of Corresponding Member (1891–95) required no payment of dues. Grade of Subscribing Member (1891–94) required annual dues of $10 for Residents and $5 for Non-Residents.

[a] Reduced dues paid by members outside the United States.

[b] Lump sum single payment.

[c] Fellowship abolished in 1950.

requirements have limited solicitation of new members, partly because of the potential for embarrassment to a candidate who might be denied admission. Because of this, ASCE did not conduct membership campaigns until 1974, and the distribution of application forms had been carefully controlled.

The 1974 policy change resulted for the first time in advertisements in the technical press inviting membership applications. More aggressive promotional effort was also encouraged in the Local Sections. The effect of this new policy on the stature and prestige of Society membership awaits future assessment.

ASCE has always been concerned with membership quality rather than numbers. Members must invest financially and in thousands of man-hours of voluntary committee service. This is the only way to provide the meaningful "professional identification" compatible with the responsibilities of the civil engineering profession.

Engineering Registration

Toward the end of the nineteenth century, it became apparent that ASCE membership was not sufficiently recognized as a mark of professional competence to protect the public from incompetent engineering. The numerous and often tragic failures of railway bridges and dams in the period 1870–95 were certainly in many cases the result of the inadequacies of unqualified persons calling themselves civil engineers.

The movement toward registration legislation originated within the profession. The reaction was so strong, however, that almost 40 years were to elapse before ASCE endorsed registration. Nevertheless, the Society played a dominant role in the development of the registration process.

At the 1897 Annual Meeting of ASCE, S.C. Thompson proposed this resolution:

> The engineering profession and the people of this country have no legal protection against any incompetent or unscrupulous person who chooses to advertise or sign using the title of Civil Engineer, thereby reflecting upon and injuring the entire profession, and
>
> In consideration of the fact that it has been found advantageous in most states to grant legal protection to members of the legal and medical professions:
>
> Therefore Be It Resolved: That the American Society of Civil Engineers places itself on record as favoring judicious legal restrictions against the unauthorized and improper use of the title of Civil Engineer.

The resolution was promptly tabled, as it was when presented again at the 1898 Annual Meeting and still again at the 1899 Annual Meeting. But Thompson's initiative and persistence brought the issue into focus. At the

1901 Convention, two similar questions were posed and were discussed to great length:

1. Do the interest of the profession, and the duty of its members to the public, require that only those who are competent be allowed to practice as Civil Engineers?
2. Under what authority, through what agency, and upon what evidence of competency should applicants be admitted to the practice of Civil Engineering?"

The discussion resulted in a proposal that licensure be administered at the state level by a Board of Examiners, with three classes of license: the first class to identify engineers of acknowledged eminence, the second class to encompass engineers of proved qualification, and the third class to include recent graduates of acceptable engineering colleges or nongraduates in their early years of practice. Advance in class of license would be by examination. The Convention attendees agreed:

> There can be no doubt that National legislation would be preferable (to many state laws) if it were possible, but it must be remembered that legislation of the character contemplated is, by the Constitution of the United States, distinctly reserved to the several states, and the Congress has no authority to enact such laws.

In 1901, ASCE appointed a Committee on Regulating the Practice of Engineering, chaired by Samuel Whinery. (Whinery was the first signer of the 1893 request from the Cincinnati Association of Engineers that the Society develop a code of ethics.) The committee submitted a report in May 1902 that described laws regulating civil engineering practice in Canada and in Mexico, and of medicine, law, pharmacy, and dentistry in the United States. The report noted laws regulating the practice of architecture enacted in California, New Jersey, and Illinois.

The committee recommended no further action by the Society at that time, giving four reasons: the difficulty of obtaining consistent state laws; the desirability that any laws cover all branches of engineering; the lack of interest in the subject outside ASCE; and the fact that, unlike physicians and lawyers, only "engineers engaged in public service" actually required regulation by law in the public interest. The Board accepted the recommendation.

The Society avoided the issue of registration at the national level during the next eight years, but this did not deter action by members in their own states. In 1907, Wyoming enacted the first state registration law for engineers in the United States. Clarence T. Johnston, State Engineer of Wyoming, expressed concern about the widespread practice of engineering by incompetents:

> Those proposing to use the waters of the state were obliged to file an application for permit in the office of the State Engineer on pre-

> scribed forms and accompanying the same with a map showing streams, canals and reservoirs and lands to be irrigated. Under the law a survey was required. However, I discovered that lawyers, notaries and others were making the maps and signing them as engineers and surveyors.

Johnston was one of 11 engineers in Wyoming enrolled in the Member grade of ASCE in 1907. In consultation with two of them—William Newbrough and Edward Gillette—Johnston drafted a bill and submitted it to the state legislature. Self-styled engineers and surveyors who had created the problem caused some delays, but the law was enacted. Johnston was chairman of the new Board of Engineering Examiners. The Board held examinations, and within two years 134 engineers were registered as Land Surveyors, Topographic Engineers, Hydraulic and Topographic Engineers, Construction and Designing Engineers, and Administrative Irrigation Engineers.

The Wyoming law set off a chain reaction. Louisiana followed with a licensure act in 1908. By 1910, the legal regulation of professional engineering practice was an issue in many states, and by 1920 12 states had enacted such laws.

When engineering licensure legislation was introduced in the New York State Assembly in 1910, the author of the bill, the Honorable J.L. Raldires, requested ASCE's support. No stated position was taken on the bill by the Society, but it was withdrawn after a delegation headed by President John A. Bensel conferred with the New York State Commission on Public Education in Albany. (Seven years later, in 1917, the minutes of an Executive Committee meeting stated that the Society "had taken action in defeating vicious legislation in New York some years ago.")

In anticipation of the consideration of further licensing legislation by the New York State Legislature, the Board of Direction adopted this resolution in April 1910:

> Resolved, that it is the sense of the Board that it is the duty of the American Society of Civil Engineers to use its influence in the proper formulation of all legislation by the general government or by any of the States of the Union, which affects the practice of Engineers and the Board recommends the appointment by the Society of a Committee whose duty it shall be to formulate the general lines on which such legislation shall be based, and that said Committee be requested to report at the next Annual Convention.

The First Model Law

In January 1911, ASCE adopted the first "model law," an "Act to Provide for the Licensing of Civil Engineers." The guide prescribed 15 sections embodying "proper requirements" to be imposed in licensing practitioners, and it recommended these minimum qualifications:

1. No less than 25 years of age.
2. Good moral character.
3. The individual shall have been engaged in Civil Engineering work, as an assistant to a licensed practitioner for at least six years, and in responsible charge of such work for at least one year.
4. For a graduate of a school of engineering of recognized reputation, the terms of actual engagement as an assistant shall be four years.

The guide further recommended that each state's licensing law should be administered by a State Board of Examiners made up of five qualified Civil Engineers, each with at least 10 years of practical experience, and that the annual registration fee not exceed $25. To achieve nationwide uniformity in these laws, the report suggested appointing three Corporate members of the Society from each state to serve on the Standing Committee on Registration of Engineers. The model law stated, in part:

> *Whereas:* There are National Societies of Engineers in the United States, membership in which can only be secured after rigid examination of the fitness of applicants to practice as Engineers; and ... the public has ample protection if they will employ only those who have thus demonstrated their ability; be it resolved that the Board of Direction of (ASCE) does not deem it necessary or desirable that Civil Engineers should be licensed in any State; and be it further
>
> *Resolved:* That if notwithstanding this, the Legislature of any State deems ... a statute covering the practice of Civil Engineering desirable for the protection of the public, the accompanying draft ... prepared by the Board ... is recommended.

Clearly, the Society became involved in the licensing process unwillingly. The American Institute of Architects, which also had opposed licensing some years earlier, was also influential in shaping the pattern of registration, while opposing it.

A special Committee on Licensing of Engineers was formed in 1911 to keep the Board of Direction informed as state after state adopted registration laws. The Society encouraged Local Sections to become involved with licensing legislation only when it became a live issue in the state.

A 1913 proposal by C.H. Higgins to put the issue of statutory regulation of engineering practice to a ballot of the membership was summarily quashed by a tabling action. The refusal to discuss the public interest aspects of the matter casts a shadow upon the image of the Society.

In 1923, the Board adopted a resolution put forth by the Committee on Licensing of Engineers. In part, the resolution stated: "Laws requiring registration or licensing of engineers and (or) land surveyors are in force in twenty-five states; and ... this Board of Direction, without endorsing the registration of professional engineers, considers that any State laws adopted should be uniform."

The President appointed a Standing Committee on Registration of Engineers, with at least one member of the Board of Direction to draft a registration law for professional engineers and land surveyors for Board approval. The committee did this, and after circulating successive drafts for several years, a satisfactory model law appeared in 1919 under the title "Recommended Uniform Registration Law for Professional Engineers and Land Surveyors." The ASCE draft law was then presented to other engineering societies in a series of special conferences, and after three more years of intersociety committee discussions and revisions, a resolution emerged.

Only three national organizations were sufficiently interested to join ASCE initially in adopting the model law: the American Society of Heating and Ventilating Engineers, the AAE, and the NCSBEE. The ASME approved "in principle." There may have been difficulty in reconciling the public works orientation of the civil engineer with the industrial orientation of other societies.

Continuing joint negotiation and amendment resulted in endorsement by 8 national organizations by 1937. Intensive efforts in 1946 brought the list to 13 societies.

ASCE issued policies on licensing in 1935, followed by a more comprehensive one in 1957. The later policy stated ASCE's support and endorsement of the registration of engineers and land surveyors as being in the best interests of the public, supported the model law and the NCSBEE, and offered specific conditions under which corporations might be registered or licensed.

Since 1960, periodic review and maintenance of the model law have been managed by the the NCSBEE's successor, the National Council of Engineering Examiners (NCEE). Although ASCE has continued to take an active part in the operation, through its participation in NCEE, conflicts with ASCE registration policy have precluded endorsement of the model law by ASCE since 1946. The ASCE position, which arises from the large segment of engineers in private practice, illustrates the difficulty of achieving absolute unity of opinion throughout the profession.

In March 1965, the Board of Direction adopted a position paper questioning the propriety of NCEE influencing the legislation that its member state boards are required to administer. Specifically, ASCE objected to efforts to integrate registration boards more closely with the state administrative structure, and to extending the registration requirement to all engineers, regardless of the basic public protection function of registration. Efforts to register corporations were also opposed. Civil engineers far outnumber any other branch of engineering in being registered. Even before registration was made a requirement for the new Fellow grade in 1959, surveys showed that about two-thirds of all ASCE members were registered in at least one state. Taking into account the foreign members not subject to U.S. registration, the overall proportion of domestic ASCE members registered and in the Engineer-in-Training phase is about 90%. By 1970, there were 55 engineering registration boards operating in the 50 states, the District of Columbia, the Commonwealth of Puerto Rico, the Panama Canal Zone, Guam, and the Virgin Islands.

Engineering registration acquired a new aspect in 1974, with proposals in some states to make license renewal contingent on the licensee keeping up with advancing technology in the field. In ASCE, the impact on continuing education activities would be especially great because between 80% and 85% of its members are registered.

Other Qualification Ideas

There have been two proposals made regarding engineering qualification that did not originate in ASCE. In 1934, ECPD proposed publishing an annual roster of engineers who could qualify by "eight or more years of active practice in engineering work satisfactory to the examining body and the passing of written and oral examinations" that might result in skills approximate to those gained from an approved engineering course. ECPD anticipated a roster of up to 100,000 names. The ASCE Board of Direction approved the plan in April 1935 after it was approved by ASME and NCSBEE. There was resistance by other ECPD societies, and in 1936 the ECPD Committee on Professional Recognition stated that "any additional procedure of certification or recognition would only be adding a fourth method to the three methods already established. It would introduce new competition or conflict and new difficulties of correlation."

The other proposal was in 1970. Noting that less than half the graduate engineers in the United States were registered, NCEE proposed that a system of "accreditation" be adopted as an adjunct to licensure. Under the plan, an engineer could attain professional status either through the conventional registration process or could become an "Accredited Engineer" through a "coordinating agency" established by the engineering societies. The idea aroused little enthusiasm.

The early lack of ASCE member interest for legal registration can be understood. The process meant relinquishing an important responsibility to a quasi-political function. Once members realized that licensure by state law was reality, ASCE became a national leader in the effort.

The progression of model laws advanced by the Society depended on administration by state boards. At ASCE's local level, Sections were aware of appointments to the state boards, and possible amendment of the laws, and were quick to react to threats of undue political influence.

Except for ASCE's conservatism, the registration movement might have gone too fast and too far in "closing" the profession as a self-serving measure. The early resistance to registration was anything but altruistic, but after 1930, the Society's main concern was the public's interest.

Ethics and Professional Practice

In the nineteenth century, ASCE concern with ethical matters was similarly plagued by ultraconservatism. Again, recalcitrance turned into leadership. By 1912, both AIEE and ASME preceded ASCE in adopting a code of ethics. ASCE's code, enacted in 1914, included a conscientious enforcement policy, which made the Society unique in the engineering world.

Informal though it might have been in the mid-1850s, the practice of civil engineering was strongly imbued with the ethic of public service necessary for those responsible for large and expensive public works. However, professional ethics were considered strictly personal—a matter of honor. John B. Jervis emphasized the engineer's relationships in a classic address at the first Annual Convention, in June 1869. He said, in part:

> The engineer eminently depends on character. The engineer has to deal with men ... with whom many questions arise that are not to be determined by a simple computation, and even computations will be questioned. With these men the engineer holds delicate relation, as the umpire between the contracting parties; and will often be laced between plausible claims on one hand, and a sense of duty on the other. ... The engineer must aim to do justice between the parties ... [and] is in duty bound to render equity, according to the terms of the contract. ... In all such cases, committed to the judgment of the engineer, he will need the best experience as a business man, and especially to cultivate the golden rule of doing as he would be done by.

The first proposal of ethical nature to come before the Society was the following resolution submitted in a meeting on October 18, 1877, by Secretary Gabriel Leverich:

> *Whereas:* A Civil Engineer, in the practice of his profession, is sometimes restrained or overruled by his employers, in matters involving serious risk to property and life which he only, as the engineer should determine; whence he must either discharge his duties in a manner contrary to his best judgment or resign his position;
>
> *Resolved:* That in the opinion of the American Society of Civil Engineers, it is unprofessional for a civil engineer to continue the discharge of his duties when so restrained or overruled; or to accept an engagement which is generally known to have been vacated because of such interference ... until the judgment of his predecessor has been formally disapproved by other disinterested civil engineers.

This resolution, and an alternately worded one offered by Past-President William J. McAlpine, were considered at the 1877 Annual Meeting and were referred to the Board of Direction. However, the Board resolved that "it is inexpedient for this Society to instruct its members as to their duties in private professional matters." The Board took no further action upon the Leverich and McAlpine motions.

Fifteen years later, in 1893, there was persistent reference in *Engineering News* to the need for a code of ethics in engineering. The issue came before the Board of Direction from the ASCE's Cincinnati Association , which stated that "the time has come when it is both possible and desirable that a Code of Ethics for the guidance of the members of the profession of Civil Engineering should be formulated and adopted."

The proposal to appoint a special committee to draft a code was presented at the 1893 Annual Meeting, but the Board did not favor such a move. The issue did not emerge again for almost a decade, except for a few papers published in *Transactions.* The topic, "The Regulation of Engineering Practice by a Code of Ethics," was a program feature of the May 1902 Convention. The consensus was expressed in an address by President Robert Moore: "And it is ... safe to say that ... the Society and its members will continue to rely, as ... in the past, upon these vital and moral forces, and not upon the enactment of codes or upon any form of legislation."

Percival M. Churchill offered the following proposal at the 1913 Annual Meeting:

> *Moved:* That this Society shall consider the adoption of the Code of Ethics for Engineers recently proposed by a committee of the American Society of Mechanical Engineers and published in *Engineering News,* January 1st, 1913.
>
> That the proposed Code be printed as a letter ballot and submitted to the members of this Society with the request that each member vote to accept or reject the Code article by article; that where a member so desires he shall—after voting against a certain article—submit a substitute in the form he desires the article to take in the same manner additions may be presented.
>
> This ballot to be closed on April 1st, 1913.
>
> The Board of Direction shall then send out a second letter ballot giving the Code as first proposed and also any suggested changes and additions. Members shall again vote article by article. The ballots to be opened at the next Annual Convention, and any article having a majority of the votes cast shall be declared adopted.

In February 1913, the Board of Direction authorized a committee "to report on the whole question as to whether it is desirable for the Society to adopt a Code of Ethics, and if they think favorably of it, to present to the Board a short Code of Ethics."

Two of the five committee members who had drafted the ASME code, Charles T. Main and Spencer Miller, were members of ASCE. The special committee, consisting of Mordecai T. Endicott, John F. Wallace, and Henry W.

Hodge, transmitted its draft of a "short" code to the December 1913 meeting of the Board. The draft was reviewed in the next two meetings, amended, and then approved for favorable recommendation to the membership at the June 1914 Convention. The draft code was adopted by the meeting and ordered submitted to the Corporate Members for mail ballot.

Code of Ethics Adopted

On September 2, 1914, this Code of Ethics was adopted by a vote of 1,997 ballots in favor out of 2,162 ballots cast:

> It shall be considered unprofessional and inconsistent with honorable and dignified bearing for any member of the American Society of Civil Engineers:
>
> 1. To act for his clients in professional matters otherwise than as a faithful agent or trustee, or to accept any remuneration other than his stated charges for services rendered his clients.
> 2. To attempt to injure falsely or maliciously, directly or indirectly, the professional reputation, prospects, or business of another Engineer.
> 3. To attempt to supplant another Engineer after definite steps have been taken toward his employment.
> 4. To compete with another Engineer for employment on the basis of professional charges, by reducing his usual charges and in this manner attempting to underbid after being informed of the charges named by another.
> 5. To review the work of another Engineer for the same client, except with the knowledge or consent of such Engineer, or unless the connection of such Engineer with the work has been terminated.
> 6. To advertise in self-laudatory language, or in any other manner derogatory to the dignity of the Profession.

1974 Codes of Ethics, as amended to October 19, 1971:

> It shall be considered unprofessional and inconsistent with honorable and dignified conduct and contrary to the public interest for any member of the American Society of Civil Engineers:
>
> 1. To act for his client or for his employer otherwise than as a faithful agent or trustee.
> 2. To accept remuneration for services rendered other than from his client or his employer.
> 3. To attempt to supplant another engineer in a particular engagement after definite steps have been taken toward his employment.
> 4. To attempt to injure, falsely or maliciously, the professional reputation, business, or employment position of another engineer.

5. To review the work of another engineer for the same client, except with the knowledge of such engineer, unless such engineer's engagement on the work which is subject to review has been terminated.

6. To advertise engineering services in self-laudatory language, or in any other manner derogatory to the dignity of the profession.

7. To use the advantages of a salaried position to compete unfairly with other engineers.

8. To exert undue influence or to offer, solicit or accept compensation for the purpose of affecting negotiations for an engineering engagement.

9. To act in any manner derogatory to the honor, integrity or dignity of the engineering profession.

The Code of Ethics has been amended eight times (1934, 1941, 1949, 1950, 1956, 1961, 1962 and 1971). The 1914 and the 1974 Codes are reproduced here for comparison. All the original articles and the preamble were modified. One article does not even appear in the 1974 version. The fourth article was changed several times from:

> 4. To compete with another Engineer for employment on the basis of professional charges, by reducing his usual charges and in this manner attempting to underbid after being informed of the charges named by another.

In 1949, this became: "To participate in competitive bidding on a price basis to secure a professional engagement."

Earlier clauses on the price of engineering services were combined into this single concise provision in 1961: "To invite or submit priced proposals under conditions that constitute price competition for professional services."

Competitive bidding was a problem as early as 1925 and became critical in 1953, when the Chief Highway Commissioner of South Carolina published a formal "Notice to Bridge Engineers," stating that sealed bids to be submitted on official proposal forms would be received for engineering services in connection with a bridge project. A number of consulting firms responded to the invitation, and the award was made to the lowest bidder. The Highway Commissioner and principals in 13 of the firms that filed bids were charged with violation of the Code of Ethics by their participation "in competitive bidding on a price basis to secure a professional engagement."

After investigation and deliberation in the Committee on Professional Conduct and the Board of Direction, with careful observation of "due process" requirements, the 14 members of the Society were found guilty of the charges. The Highway Commissioner was expelled from membership, one consulting engineer was suspended for five years, and the remaining 12 bidders were suspended for one year.

This landmark case began an era of great effort by ASCE—later joined by the NSPE and the Consulting Engineers Council—to encourage prospective clients to engage professional engineering services by the process of "professional negotiation." In this procedure, only the quality of the service is considered, and competition on a price basis is precluded. ASCE waged its campaign with manuals and pamphlets, hundreds of letters, telegrams, and telephone calls, and through staff contacts with public- and private-enterprise agencies employing consulting engineers. The total effort was educational, but it emphasized that ASCE members who engaged in competitive bidding were acting in an unethical and unprofessional manner. There were occasional professional conduct proceedings for violations of the competitive bidding provisions, but none attained the importance of the 1953–54 "South Carolina Case."

This effort from 1954 to 1971 resulted in almost universal acceptance of "professional negotiation" among the users of professional engineering services. Early in 1971, however, as a result of insistence on competitive bidding by certain federal officials, the U.S. Department of Justice charged the Society with violation of the antitrust provisions of the Sherman Act. The charges centered on the "Article 3" competitive bidding clause in the Code of Ethics and on its administration.

The action was not altogether unexpected. In 1956, an agent of the Department of Justice visited ASCE headquarters and investigated in detail the Society's ethical standards and activities relating to professional services engagement. There was no official communication after that visit. But in 1961, ASCE's counsel warned that the administration of the competitive bidding policy as an ethical matter was vulnerable to Sherman law interpretation. This was followed soon after by Department of Justice inquiries concerning similar competitive bidding provisions in the professional practice codes of the American Institute of Certified Public Accountants and the American Institute of Architects.

In 1971 the Board of Direction, under the leadership of President Oscar S. Bray, deliberated the advice of eminent legal counsel as to the chances for successful defense of the suit. Further, legal counsel informed the Board of the possibility of dire restrictive and fiscal consequences of an unsuccessful defense. On the advice of counsel, the Board considered a negotiated settlement that would permit the Society to continue to advocate professional negotiation by education, persuasion, and legislation, if not as a matter of ethics.

The Board acted with considerable courage and conviction in October 1971 when it moved voluntarily to delete Article 3—"[t]o invite or submit priced proposals under conditions that constitute price competition for professional services"—from the Code of Ethics. About 500 members (of 67,000) reacted with letters, telegrams, and telephone calls protesting the deletion. They were largely mollified when the circumstances were explained.

The voluntary elimination of the ethical standard in question opened the door to a Consent Decree settlement of the Department of Justice suit in mid-1972. The settlement left essentially intact the vitally important Manual of

Practice No. 45, *Consulting Engineering—A Guide for the Engagement of Engineering Services*, which sets forth the traditional ASCE professional negotiation procedure recommended for so many years. Ironically, only a few months after the suit was settled, Congress enacted legislation—with vigorous ASCE support—prescribing a professional selection and negotiation procedure for engagement of architect engineer services (Public Law 92-582, 92d Congress).

A footnote added to the Code of Ethics in October 1963 is an effort to enable American engineering firms engaged in foreign work to compete with firms from countries in which there are no ethical constraints. The provision, promptly dubbed the "When in Rome Clause," has been highly controversial; several efforts to invalidate it were narrowly averted.

Professional Practice Guidance

Apparently, the "short" Code of Ethics issued by the Board of Direction in 1913 permitted some latitude of interpretation. In about 1925, a committee of the Northeastern Section of the Society presented a "Code of Practice" primarily "for those engaged in construction," which provided practical guidance to the individual engineer. The Board of Direction approved the document after revision by the Committee on Professional Conduct. It was published in 1927 as Manual of Engineering Practice No. 1, initiating a distinguished series of ASCE publications.

Daniel W. Mead, of the University of Wisconsin, was an authority on ethics in the history of engineering. During his tenure as ASCE President, he observed in 1936 that "most engineering ethical codes seemed to apply almost exclusively to the engineer in general practice and not to the more than 90% of the profession who are public or private employees." He took this view to the Committee on Professional Conduct and the Board of Direction, whose members decided the word "client" in the Code of Ethics was to be inclusive of the term "employer." They invited Professor Mead to prepare "an inclusive paper on this subject as a basis for a wide and full discussion, and that such paper after discussion be made readily available to all engineers." The result was Manual of Engineering Practice No. 21, *Standards of Professional Relations and Conduct*, published in October 1940. Thousands of copies of this classic monograph have been distributed since.

In 1961, ASCE published an official "Guide to Professional Practice Under the Code of Ethics" as an adjunct to the Code of Ethics.

ASCE has supported ethical standards other than its own. ECPD adopted its Canons of Ethics in 1947; ASCE became a signatory to those canons in January 1950, but in 1964, ECPD made certain amendments. ASCE withheld formal action on these on the grounds that there might be ambiguity in the interpretation of the Society's own Code of Ethics.

There have been attempts at a single code of engineering ethics that might be acceptable to all engineering societies. The unsuccessful effort in

1913 to approve a code drafted by ASME was the first. In January 1921, the four Founder Societies, together with the American Society of Heating and Ventilating Engineers, formed a joint Committee on a Proposed Universal Code of Ethics that recommended a code for universal adoption. They also recommended that each society appoint a Committee on Professional Conduct. ASCE's Board of Direction rejected the proposed universal code. The Board favored the idea of a Committee on Professional Conduct, which would afford more than a guilty or not-guilty judgment of unprofessional actions, and it created the standing Committee on Professional Conduct that still exists.

Efforts to make the ethical standards of ASCE and the NSPE compatible have met with only limited success. The ethical emphasis for those engaged in manufacturing, such as mechanical, electrical, and mining engineering, differs from the attitudes of those engaged primarily in the public works sector. The ASCE membership also includes many more engineers in private practice than is the case with other societies, and this is reflected in ASCE ethical attitudes.

ASCE's Code of Ethics serves as a pattern for several other domestic societies and registration boards, and it has had a significant impact on foreign ethical standards. The influence of ASCE's Code is evident in the codes adopted by the Pan American Federation of Engineering Societies (UPADI) in 1961, the Conference of Engineering Societies of Western Europe and the U.S.A. (known as EUSEC) in 1963, and the World Federation of Engineering Organizations (WFEO) in 1969.

Ethical standards have meaning only if they are administered and enforced with determination, fairness, and consistency. ASCE's record in its enforcement of its Code of Ethics is unequalled by any other engineering organization in the world.

The 1852 Constitution and those since require members of the Society at the time of admission to certify by signature subscription to the Constitution and Bylaws. Since 1878, the Constitution has provided for the expulsion of members for cause. The first recorded cases were two in 1899; in one the charges were dropped, and in the other a member was requested to resign.

The Board of Direction dealt directly with all disciplinary matters until 1923. Since then, the Committee on Professional Conduct advises the Board on those matters of ethical nature referred to it. Until 1968, the Committee on Professional Conduct was a committee of the Board; since that time, only past Board members sit so that no current members of the Board function both as prosecutor and as judge in any proceeding. Committee duties include the investigation and judicial review of all charges of unprofessional conduct as a basis for possible disciplinary action by the Board.

The Society's records show that, between 1876 and 1900, two cases were filed. In one case, the charges were dropped; in the other, the member's resignation was requested. Between 1900 and 1925, 40 cases were filed and charges were dropped in 31 of them. The number of cases heard steadily increased, and between 1951 and 1974, 185 charges were filed with the Committee. Of those, 26 members were admonished, 25 were suspended from ASCE, four were asked to resign, and 23 were expelled. Charges were dropped in the remaining 130 cases. This is a record unique in the engineering profession, both for the number of cases and the manner of their disposition.

In addition to the "South Carolina Case," three other cases merit brief comment here because of their unusual nature.

In the late 1920s, a flood control district in a Western state began constructing a dam, and it abandoned the project before completion because of foundation problems. There were accusations of irregularities, and a member of the Society was highly critical in his public expressions on the matter. Some of his comments reflected on the professional reputation of the chief engineer of the flood control district, who had been exonerated from any connection with the alleged irregularities. After an investigation and hearing, all in strict accordance with the rules of the Society, the critical individual was expelled from membership, with considerable attention on the part of the public press.

There were four efforts to induce the Board of Direction to rescind the expulsion, none presenting additional evidence. The last request followed the death of the member, 35 years after the expulsion occurred.

Another case, in the 1960s, involved a young member employed by a consulting firm. He was charged with unprofessional conduct for beginning his own private practice before terminating his employment. When the matter came to the point of a hearing before the Board, the defendant secured an injunction against such a procedure. At considerable expense, the Society had the injunction set aside, and the hearing went forward some months later. The defendant was found guilty of the charges against him, and he was suspended from membership for five years. The cost of his case to the Society for legal services and related expenses was more than $20,000.

The most spectacular ASCE professional conduct case resulted from charges against three members of the Society involved in political influence-peddling activities in Maryland. These activities resulted in the resignation in 1973 of Spiro T. Agnew from the Vice Presidency of the United States. Two of the defendants represented consulting firms who made contributions in return for favored treatment in the award of engineering contracts; the third, as a public employee, was a go-between in arranging the payments. All were charged with violating the ethical provision against undue influence or the offering, solicitation, or acceptance of compensation "for the purpose of affecting negotiations for an engineering engagement."

Even though the defendants had gained immunity from criminal action by providing evidence of the Maryland corruption, they were found guilty and expelled from membership in the Society in 1974. They had been given full opportunity to answer the charges in writing and in a hearing before the Board of Direction. It appeared likely at the time that other members might also face action by ASCE.

When Past-President and Honorary Member Daniel W. Mead criticized the ASCE ethical standards in 1936 for their emphasis on the engineer in private practice, he sounded a note that echoed long after. Other critics have observed that the Code focused on "business ethics "at the expense of "personal ethics." The Environmental Movement of the 1960s brought allegations that engineers were subordinating the "social ethic," under which they were obliged to accept responsibility for any negative impact of their works upon man and his surroundings.

These criticisms were valid. Certainly, the 1974 ASCE Code of Ethics does not call for consciousness of human relationships and social responsibility. However, the engineering profession alone is not at fault for all the problems that have befallen society with the progress of technology. The civil engineer's concern for the public welfare is implicit in the general provisions of the Society's ethical standards. From all indications in 1974, the objectives of the Society were ranging more and more into the social concerns of civil engineering practice.

CHAPTER 5

Creating the Professional Environment

A profession cannot limit its concern to the education, competence, and practice of its members. To be effective, the professional must operate in a favorable milieu. Adequate compensation, appropriate recognition by one's employer, provision for professional development, authority to exercise judgment, and freedom from undue political or bureaucratic constraints are some essentials of that milieu. The professional society provides the mechanism for any collective action in such matters. ASCE has served where it could, while avoiding the attitudes and methods of trade unions. This chapter addresses ASCE's efforts to enhance the economic welfare of civil engineers, young engineers, and women members of the Society, and to encourage professional development.

Employment Categories

Civil engineers work in the public, private, and industrial sectors. Data compiled in 1964 (Table 5-1) show the member employment in several professional organizations. More civil engineers than other engineering professionals practice in the public works sector. Other engineers work mainly in manufacturing. Only about one in eight of all civil engineers is employed in an industry other than construction. Civil engineering also has strong representation in the private practice sector, although not as much as in architecture, law, and medicine. The trend in civil engineering, at least since 1900, has been from government service toward private practice. This analysis is significant because it explains the marked differences in professional objectives and

Table 5-1. *Members' Areas of Practice in Engineering and Other Professional Organizations, by Percentage of Membership, 1964*

Area	*ASCE*	*AIME*	*ASME*	*IEEE*	*AIA*	*ABA*	*AMA*
Private practice (principals and employees)	29	9	10	10	68	76	63
Government (federal, state, and local)	40	4	9	9			
Industry (including construction)	23	76	73	68			
All other	8	11	8	13			
Government, industry, and all other combined					32	24	37

Note: AIME is American Institute of Mining Engineers; ASME is American Society of Mechanical Engineers; IEEE is Institute of Electrical and Electronics Engineers; AIA is American Institute of Architects; ABA is American Bar Association; AMA is American Medical Association.

attitudes on the part of the civil engineer, as compared with industry-oriented engineers.

Polarization between practitioners in government service, as represented by the U.S. Army Corps of Engineers, and nonmilitary civil engineers was probably taking place at the time ASCE was being formed. As early as 1870, there were differences in judgment and opinion between the engineers of the Corps, usually graduates of West Point, and those of nonmilitary origin. The problem surfaced at the Saint Louis Convention in 1880, when Charles MacDonald (President, 1908) bitterly protested the education

> at public expense, of a privileged class of engineers in military service, to whom are entrusted the design and supervision of works of public improvements, to the exclusion and prejudice of engineers in the civil service, whose education has not been a tax on the public treasury, who have proved themselves perfectly competent to execute works of the greatest magnitude in the best and most efficient manner.

His pleas resulted in appointment of a Committee on the Engagement of Civil Engineers on Government Works, "to prepare a memorial to Congress asking that Civil Engineers may be placed in full charge of the public improvement carried on at Government expense." With MacDonald serving as chairman, the committee brought its draft "memorial" before the 1881 Annual Meeting, with the admonition that

> it may be inexpedient for the Society to place itself in the position of advocating before Congress the claims of a certain class of its membership, in seeming conflict with any other class whose interest may be in a different direction.

The memorial was submitted for the use of any civil engineers who might wish to endorse it as a personal communication to Congress, with the suggestion that "the Society may with propriety decline to consider the subject further." The issue arose again just five years later when the Civil Engineers Club of Cleveland sought ASCE cooperation in exploring the desirability "of new ideas which shall provide for the better condition of Civil Engineers employed on Government works other than military." The Society declined to participate, citing the conclusions reached in 1880.

The dichotomy became an open controversy in 1885 in the paper, "Ten Years Practical Teachings in River and Harbor Hydraulics," presented by Elmer L. Corthell (President, 1916). He maintained that civilian engineers were better qualified by education and experience than military engineers to direct major public works projects. Arthur P. Boller (Secretary, 1870–71; Vice President, 1911–12) offered a compromise view in his observation that there were well-trained and competent engineers both within and outside the military, and he urged that the specialist services of capable civilian engineers be used in the direction of government projects. Actually, by this time the influence of West Point as a source of planners and builders of major public works had already begun to decline.

The issue did not go away. In 1920, Charles MacDonald headed a group "with the object of urging suitable recognition of Civilian Engineers employed on River and Harbor Works, in the proposed legislation providing for an increase of the Engineer Corps of the U.S. Army." This effort was successful. The 1914 legislation included a provision requiring that all civilian assistant engineers engaged in the Engineer Bureau of the War Department have at least five years experience, and that they be given the rank of captain or major in the Corps.

The controversy emerged again in 1921. A Board of Direction resolution objected to the lack of consideration given by the War Department to civilian engineers who had been given responsibility over important river and harbor works during World War I. When the war ended, the civilians were replaced in their supervisory posts by commissioned officers and were demoted to their former duties. Congress was urged to enact legislation that would afford protective status to civilians under a new classification of "United States Engineer." The resolution drew fire from General Lansing Beach, M.ASCE, Chief of the Corps of Engineers, who considered it a reflection on his office. There was no other apparent result. The last confrontation between the civilian and military came in 1924, when the Society's supported legislation that would have placed responsibility for all nonmilitary federal public works in a new department, outside the War Department. This time General Beach appeared before the Board of Direction to voice his objections. The legislation was not enacted.

The Corps of Engineers welcomed the aid of the Society in 1949 when Congress considered a drastic reorganization of the Army. The proposal would have abolished the existing technical services (Corps of Engineers, Signal Corps, Quartermaster Corps, Transportation Corps, and Ordnance Department), and it would have assigned any or all duties of these branches to the general branches such as Infantry and Field Artillery. ASCE, through its Committee on Military Affairs and through the Engineers Joint Council (EJC), supported "the retention and further development of the technical and professional branches of the Army, and the maintenance of professional status to a high degree among the officers of the technical branches of the Army."

After World War II, any schism between the military and civilian interests appeared to vanish. The relationship between the Society and the civil engineering units of the Army, Navy, and Air Corps was most cordial.

The Great Depression created contention between engineers in the private and public sectors regarding the growing bureaucracy in all government agencies involved with public works. This included such bodies as the Bureau of Reclamation, Bureau of Public Roads, Geological Survey, Coast and Geodetic Survey, Corps of Engineers and, later, such new agencies as the Atomic Energy Commission and National Aeronautics and Space Agency.

A 1938 resolution drafted by the Committee on Fees and approved by the Board of Direction expressed concern about the encroachment of organizations into fields formerly occupied by engineers in private practice. The resolution disapproved all proposals that would unduly restrict activities of the engineers engaged in private practice, and it particularly disapproved of any laws that might deprive the public of benefits derived from the services of the private engineer. The Society also reaffirmed safeguards for tenure under civil service for engineers in the service of the public. In 1941, the Board recorded its disapproval of "the accelerated trend toward socialization" of the engineering profession, and it urged the abandonment of efforts by federal agencies to perform "in-house" the services normally furnished by engineers in private practice. In 1945, ASCE created a Committee on Engineers in Private Practice to represent the interests of that segment of the membership. The committee promptly began discussions with the Secretary of the Interior about the Bureau of Reclamation. The discussions led to policy decisions in 1946 that were satisfactory to both groups.

In January 1948, the Board of Direction adopted a policy statement, reaffirmed in 1959, and still in force in 1974. The statement encouraged the use of private engineering consultants by government when this would best serve the public, and it discouraged unfair competition by engineers in both groups. The statement also recommended that administrators with suitable technical training head public agencies employing large numbers of engineers.

In 1963, Congress considered legislation that would have barred a government agency from "hiring" another agency to perform engineering services. These included mapping services in Ethiopia by the Army Map Service, hydrographic studies by the Coast and Geodetic Survey for the Bureau of Reclamation, and efforts by the latter Bureau to extend its services to local irrigation districts. The legislation was not enacted. By 1974, the agencies of federal, state, and local governments had become by far the greatest users of the professional services of engineering firms in private practice. The Committee on Engineers in Private Practice was merged into a Committee on Professional Practice when the Department of Conditions of Practice was created in 1953.

In 1957, the engineers in government were given their day when a Committee on Engineers in Public Practice was created by the Board of Direction. For 13 years, the committee gave attention to the status of the engineer in federal Civil Service, and to grade classifications and salary schedules for engineering employees in state and local agencies. The organizational merry-go-round came full circle in 1970, however, when the responsibilities of the Committee on Public Practice were delegated to the new Professional Practice Division. The conditions of practice for all members of the society—regardless of the manner of their employment—were again represented by the same group of committees.

ASCE's Board of Direction has not represented the segments of civil engineering numerically. Although 40% of the membership was engaged in some level of government practice, this segment had never been fully represented on the Board. This reflects the constraints that have limited participation in professional activities by many engineers in public employment. Some public bureaus openly discouraged such activities; others charged the time required against annual leave allowances.

Seeking the Members' Economic Welfare

From its beginning, ASCE has faced the dilemma of deciding the extent to which it considered the economic self-interest of its members. The practitioner should be accorded suitable surroundings and facilities with which to work, should be given the recognition appropriate to a professional; and should be reimbursed at levels commensurate with the worth of his or her services to society.

Few members of ASCE have felt the Society should fulfill the same role as a trade union. But some questioned how far the Society should go in serving the economic needs of its members, while retaining its identity as a professional body.

The founders and early members of ASCE were mature, well-established engineers, enjoying top incomes in industry and government. In the years 1850–75, civil engineers were very well compensated indeed in comparison with others, and many leaders of the Society at this time were independently wealthy. An 1875 proposal that a "Civil Engineers' Insurance League" be set up to provide benefits to families of deceased members drew little interest and was abandoned.

A letter to the membership in 1872 invited "those seeking engagement and those requiring engineering service" to communicate their needs to the Secretary, so that "such may be brought together." Any such services were provided informally by the secretariat.

The practice of civil engineering changed at the turn of the century, as the membership changed from those mostly in executive management positions to those who were employees. Percival M. Churchill presented a proposal in 1912, addressing the problem. He suggested

> that the President appoint a committee of eight members to look into the conditions of employment of Civil Engineers throughout the country, the compensation they receive, the duration of employment, the expenses for which they are reimbursed by the employer, the expenses due to the work paid by the engineers themselves, the net yearly income, the prices charged for different classes of private work, and any other facts necessary to clearly set forth the problem. The report to set forth recommendations for action by the Society looking forward to improving existing conditions and to include a report on the feasibility of this Society operating an employment bureau for its members.

The resolution was referred to the Board of Direction, which appointed a Special Committee to Investigate Conditions of Employment of, and Compensation of, Civil Engineers. The Board observed "that it does not feel that it would be practicable or wise for the Society to operate an Employment Bureau for its members." Slightly more than half of the Society's 6,805 members at this time were in the Associate Member and Junior grades.

Churchill persisted in his plea for an "Exchange for the Marketing of Engineering Service" at the 1913 Annual Meeting. This time he urged that ASCE seek the cooperation of the American Institute of Consulting Engineers (AICE), ASME, AIEE, and others in exploring such a venture. There was no action.

The Special Committee made its final report to the Board at the 1917 Annual Meeting. Although the report was an excellent compilation of data and information on compensation and employment conditions, it suggested no Society action. The average yearly compensation of 6,358 civil engineers was $3,985. Members averaged $4,141, compared with $3,389 for nonmembers. Interestingly, graduate engineers averaged $3,982 against $3,993 for nongraduates.

Churchill was not content with the report, and he filed another resolution at the 1917 Annual Meeting requesting appointment of a committee "to formulate a plan for the systematic marketing of Engineering Services." The committee was to be given a budget of $5,000, and empowered to employ a "competent progressive Engineer" to work out details of such a plan in cooperation with other national engineering societies. His motion was supported by this statement: "The energies and resources of the Society must be directed in other channels (other than technical) ... along the business and human sides of engineering activity." This time he prevailed, and the Board appointed a new committee to study the advisability of an employment board.

The committee found "a widespread demand for action by the Societies in this matter" and "a general feeling that our Society should, in the future, give systematic study to the practical matter of employment." It recommended a cooperative engineering employment agency, separate from yet supported and controlled by the societies, and proposed that it be set up.

The Engineering Societies Employment Bureau was finally established late in 1918, under the direction of the Engineering Council, but under the management of the Committee on Engineering Employment, comprising the four Founder Society secretaries. The precarious but useful life of this agency until its demise in 1965 is described in Chapter 8, which deals with intersociety operations. Though Churchill's ideas were not realized exactly as he envisioned, they were responsible for an important and productive joint enterprise.

After a year's study by a Committee to Investigate the Desirability of a Benevolent Fund, the plan was put before the membership in 1921 and passed into oblivion to join the "Insurance League" proposed in 1875.

The 1921 controversy concerning civilian engineers in the Corps of Engineers resulted in the Committee on the Status of the Civil Engineer in Government Work and His Compensation. The committee reported "progress" five times in the next 18 months, but it was discharged in 1923 without having made any recommendation of record.

After its 1917 study, the next ASCE action on engineering salaries was collaboration in and endorsement of the excellent 1920 report of the Engineering Council Committee on Classification and Compensation of Engineers. This committee was chaired by Arthur S. Tuttle, Hon.M.ASCE (President, 1935).

Salary Studies

As a result of engineering salary problems in government agencies across the country, the Board created the Committee on Engineering Employment in Public and Quasi-Public Offices in April 1927. The Society also authorized a delegation to call on the Mayor of New York City on behalf of 3,600 engineering employees; salary increases aggregating $2.5 million were granted.

The committee was renamed the Committee on Salaries in 1930, beginning a 27-year era of great productivity. The onset of the Great Depression in 1929 demanded the full attention of the committee in the early 1930s, when it had to take emergency measures to deal with severe unemployment and economic distress.

In 1931, the Committee completed a special report on unemployment under E.P. Goodrich. The committee recommended that Local Sections set up committees to gather funds for engineers in need and to help find jobs for them. At least 15% of civil engineers were unemployed, and at least 10% of these were in straitened circumstances.

This plan was implemented in large cities across the country with great success. Local Sections of other societies joined in the good work. The Professional Engineers Committee on Unemployment in New York City collected more than $100,000 and generated more than $300,000 in wages, while finding jobs for 1,389 unemployed engineers. Herbert Hoover, Hon.M.ASCE, then President of United States, contributed $5,000 to the aid of engineers. In 1932, 41 ASCE Local Sections were involved in this effort.

Until 1934, the survey data gathered by the Committee on Salaries was freely disseminated to members and to engineering employers for guideline purposes. The first official issue of a recommended schedule of position classifications, with appropriate salary ranges for each of the six classifications, was endorsed by the Board of Direction and published in 1934 under the title "Prevailing Salaries of Civil Engineers." Reprinted in 1935 as a separate report, this document found wide use, and it was adopted by several federal agencies employing engineers. Significantly, the federal Civil Works Agency used the report as the basis for correcting inequities in the compensation of engineers under the Emergency Relief Act.

In 1935, the Committee on Salaries had some relief from the welfare stresses of the Depression and resumed its salary investigations. A report was published in 1936, and in March 1940, the committee produced a "Grading Plan and Compensation Schedule for Civil Engineers." This document put forth, under the imprimatur of the Society, recommended salary schedules

> intended to be used as standards against which to measure salaries now being paid, and to be used with judgment as to any need for differentials by reason of higher costs of living and higher general salary levels prevailing in any given city or geographic area.

ASCE also sponsored an intersociety "Joint Conference on Engineering Salaries" in 1940 and undertook aggressive membership welfare services between 1940 and 1949. There was a near crisis in several state highway departments and some large city engineering departments as a result of low salary budgets, threatened unionization, and competition from nongovernment employers. When these agencies sought ASCE's assistance, the Society devised a program under which a staff specialist worked in the field at the agencies, studied job classifications and salaries, and helped develop specific recommendations for salary adjustments. This service was extended on request of the Local Sections and with the consent of the agencies, which also bore the subsistence expense of the staff expert.

This service was extended to the Arizona and Nevada state highway departments in 1940, and to the Nebraska State Highway Department in 1941. The North Dakota State Highway Department used those studies in 1941 to update its salary schedules. These activities were reported in detail in ASCE Manual of Engineering Practice No. 24, *Surveys of Highway Engineering Positions and Salaries*, published in July 1941.

In 1944, ASCE created the standing Committee on Employment Conditions. An updated "Classification and Compensation Plan for Civil Engineering Positions" was published by the Committee on Salaries. Staff consulted with five municipal departments in Milwaukee, followed in 1945 by similar work for Los Angeles County, the Maryland State Employees Standard Salary Board, and the Louisiana Department of Highways. The committee conducted a special survey of qualifications, responsibilities, and salaries of civil engineering teachers in 1947.

A new edition of the 1944 classification and compensation plan was published in 1946, and the final staff field consultation was completed in 1948 for the Office of Personnel, Commonwealth of Puerto Rico. ASCE then discontinued the service, stating that the staff salary analyses and consultations transgressed the field of private practice of management consultants.

Beginning in 1951, the Committee on Salaries produced a series of salary survey reports that carried only statistical data on prevailing salaries, with no recommendations concerning job classification and appropriate salary levels. The change in policy was accompanied by publication of Manual of Engineering Practice No. 30, *Job Evaluation and Salary Surveys*, providing guidance for the conduct of salary surveys at the local level. This manual enabled a local employer to analyze problems in compensation of engineering personnel without having to call for outside assistance.

The national surveys continued at two- or three-year intervals, although in 1957 the more broadly based Committee on Employment Conditions assumed the functions of the Committee on Salaries. In 1956, ASCE began a new service with the "Engineering Salary Index," published quarterly in *Civil Engineering*. The index reports regional changes in salaries, allowing for comparisons.

Year-to-year comparison of the results of the salary surveys is complicated by the wide variation in their format. The data in Table 5-2, which are reasonably compatible, illustrate some trends.

From 1959 to 1971, the ASCE reports included data from the surveys of income of engineering educators as made by EJC. In 1968, the mean salary at the rank of professor was $17,763; the mean professional income from all sources was $22,277.

Unions and Collective Bargaining

The 1935 National Labor Relations (Wagner) Act had an impact on the engineering profession. Wage rates for skilled and unskilled labor rose to levels incompatible with professional engineering salaries, especially for young engineers. The original Wagner Act made it readily possible for professionals to be drawn into unions whether or not they so chose, and the unions took full advantage of this loophole in the law. A few engineers favored the union route to immediate income benefits. Membership pressures called for ASCE action.

Table 5-2. *Median Annual Salary of ASCE Members, Selected Years*

Year	*Median Annual Salaries (dollars)* *Starting Rates for Graduates*	*Professional Grades*
1930	1,824	
1940	1,856[a]	3,721[b]–5,503[c]
1951	3,150	
1961	5,550	8,223
1971	9,450	16,032

[a]Average for Juniors.
[b]Average for Associate Members.
[c]Average for Members.

A Committee on Unionization of the Engineering Profession was formed in 1937, and it produced a report published in *Civil Engineering* in March 1938, summarizing the status of unionization of engineers. The report concluded that employers would have to provide equitable remuneration and working conditions for their engineering employees if collective action through the unions was to be avoided. Some of its recommendations were:

1. Trade union membership is a matter of personal economic determination, and should have no bearing on a member's status in the Society.
2. ASCE should not endeavor to bring about amendment of the Wagner Act to exclude professionals, but should support amendments to clarify the position of professionals and subprofessionals under the Act.
3. When necessary, ASCE should cooperate with other professional bodies to establish agencies affording a dignified means of collective representation for engineers.
4. The Society should "seek actively" to encourage acceptance and implementation of its recommended classification of professional employment grades and salary schedule, and should cite any member for unethical conduct who does not comply with these provisions.

In 1944, a standing Committee on Employment Conditions assumed the charge of the Unionization Committee, giving top priority to a problem given little attention by other engineering societies. The committee evolved a plan in 1943–44 under which the Local Sections might initiate the formation of professional collective bargaining groups. Manual of Engineering Practice No. 26, *The Engineer and Collective Bargaining*, provided guidance. It was obvious, however, that the only relief was amendment of the Wagner Act.

When the 80th Congress undertook overhauling the Wagner Act in 1946, ASCE pressed aggressively for incorporation of three provisions into appropriate amendments:

1. Any group of professional employees, who have a community of interest and who wish to bargain collectively, should be guaranteed the right to

form and administer their own bargaining unit and be permitted free choice of their representatives to negotiate with their employer.

2. No professional employee, or group of employees, desiring to undertake collective bargaining with an employer, should be forced to affiliate with, or become members of, any bargaining group which includes nonprofessional employees, or to submit to representation by such a group or its designated agents.
3. No professional employee should be forced, against that individual's desires, to join any organization as a condition of employment, or to sacrifice the right to individual personal relations with an employer in matters of employment conditions.

Other engineering societies were aware of the attention ASCE had given the unionization problem since 1937. EJC adopted the above three-point ASCE policy, and it set up a Labor Legislation Panel. The National Society of Professional Engineers (NSPE) and the American Society of Engineering Education added their support, and a strong presentation was made to Congress. This effort brought about enactment of the "professional employee" provisions in Public Law 101, 80th Congress, the Labor Management Relations Act, better known as the Taft-Hartley Act.

The new law recognized the differences in basic objectives and interests of professional and nonprofessional employees, and that it was not "appropriate" to include both in the same collective bargaining group. The law provided the needed statutory definition of a professional employee, and it afforded professionals the right to form collective bargaining units of their own. Professional employees were given the right to determine their course of action.

The labor lobby resisted the professional employee amendments, and it tried to effect repeal. The Labor Relations Panel of EJC, with ASCE representation, was a major factor in the defense.

Under the Taft-Hartley Act, ASCE is free to inform its members about labor relations, and to adopt policies on the subject, but action as to forming collective bargaining groups is open to challenge. The Society may not provide assistance of any kind to units under the act that must be restricted in membership to employees alone.

ASCE has monitored the progress of unionization in the civil engineering profession through Committee on Employment Conditions surveys conducted every four or five years. According to data collected by the Committee on Employment Conditions, in 1973, 4.6% of the membership belonged to a collective bargaining group. By 1973, 27% responding to the committee favored collective bargaining. Those who would join a collective bargaining group voluntarily were 24%, and those who would join such a group to hold their jobs were 39%.

The periodic surveys of employment conditions resulted in member feedback. The 1968 survey elicited many letters and comments chiding the Society for its lack of attention to engineering salaries as compared with the soaring wage rates of skilled and unskilled labor. In reporting this groundswell of opinion, the Executive Director stated: "Professionalism in engineering can survive only in a realistic and equitable economic environment."

Positive response by the Society was soon forthcoming. By 1973, a manual, entitled *Guidelines to Professional Employment for Engineers and Scientists*, was endorsed by 16 national organizations. This was supplemented by a more specific ASCE "Guideline: Employer Engineer Relationship" for civil engineers and their employers. ASCE also produced a recommended salary guide similar to those issued with Board of Direction sanction in the early 1940s. A specialist assigned to work with Local Sections on programs for improving employment conditions joined headquarters staff. Efforts were under way to develop a professionwide pension plan that would afford portability of pension privileges and equities as well as other benefits.

Professional Services Fees

As far back as 1874, Past-President William J. McAlpine had affirmed "the propriety of engineers making charges, the same as other professional men, for advice given." Consultation services at first were provided on an individual basis, but in time engineers worked in private practice firms headed by an individual or operated as partnerships or corporations. By 1974 almost a third of all civil engineers worked in consulting firms either as principals or employees.

The study of compensation of civil engineers reported in 1916 did not cover consulting fees, but it referred to them as a source of compensation problems in the profession:

> Engineers in private practice sometimes employ engineers with extensive experience, and presumably of good ability, at salaries which young graduates with little or no experience are able to command, but which are less than those of an ordinary mechanic who has labor organization support. ... The competition for work on the part of engineers who may employ a technical staff is so keen that it is necessary to take advantage of the needs of those seeking employment in order to secure professional work ... frequently let to the lowest bidder.

ASCE has long been concerned with procedures for professional service contract negotiation and the manner in which such fees are determined and charged. Fees must be adequate to ensure the highest quality of service as well as compensation to engineers commensurate with the responsibility they are expected to assume.

One of the earliest official reviews of the cost of engineering services was made in 1920 by R.L. Parsons, in a special investigation for the Engineering

Council's Committee on Classification and Compensation of Engineers. He concluded that engineering costs would amount to about 5% for railroad work, and about 7.5% for water works and municipal engineering.

ASCE, in 1927, authorized a special Committee on Charges and Methods of Making Charges for Professional Services. Its extraordinarily comprehensive report, published in *Proceedings*, September 1929, was to be the forerunner of an important series of ASCE Manuals of Professional Practice. The 1929 edition covered such matters as engineer–architect relationships, engineer–contractor relationships, and recommended fees. For general engineering services, exclusive of resident supervision, fees ranged from 9.5% for a net project construction cost of $25,000 to 4.25% for one of $2 million. Again, in 1938, ASCE issued a schedule of approved fees for structural and foundation engineers engaged by the U.S. Housing Authority and the Federal Housing Administration. The Society cooperated with the American Institute of Architects, the American Engineering Council, and ASME on these fees.

The Committee on Fees provided guidance on fees during the years 1930–45 and was succeeded by the Committee on Private Engineering Practice until 1950. Since that time, this function has been served by the Committee on Professional Practice. Manual of Engineering Practice No. 29, *Manual of Professional Practice for Civil Engineers*, was issued in 1952; Manual No. 38, *Private Practice of Civil Engineering*, in 1959; and Manual No. 45, *Consulting Engineering—A Guide for the Engagement of Engineering Services*, in 1968. The data in these publications were presented as prevailing rates, not as recommended schedules, to avoid questions from the Department of Justice with regard to price fixing. Manual No. 45 became especially important in 1972 when the competitive bidding clause was removed from the Code of Ethics by agreement with the Department of Justice (see Chapter 4). With minor revisions, the "Procedure for the Selection of the Engineer" remained intact. This section covers the professional negotiation process fundamental to ASCE policy for many years.

On several occasions, the Society has waived dues. This was done first in 1918, for about 1, 100 members (15% of the total) during the duration of their military or other war-related service. During the Depression, ASCE waived dues for more than 2,000 members in 1933 and 1934. Annual dues were deferred in 1942–45 for certain members in military service in World War II and in the subsequent involvements in Korea and Vietnam. More than 5,000 members were in the uniformed military services in 1943, about 26% of the total membership.

Group Insurance Programs

There was little interest in "Civil Engineers Insurance League," proposed in 1875, and the Benevolent Fund for "necessitous" members, proposed in 1921. However, members welcomed a 1949 plan for the Society to sponsor group disability income insurance. The program has operated continuously ever since under the direction of a professional administrator, with about 3,000 members enrolled in 1974.

The success of the Disability Income Plan led to other group insurance programs. In 1955, ASCE offered hospital and medical insurance; in 1960, life insurance; in 1961, accident insurance; in 1968, supplemental hospital insurance; and in 1972, catastrophic health and accident insurance.

The Society intended that, except for the administrator's fee, the entire group insurance premium should be used for the purchase of benefits to the policyholder. The programs were periodically modified and updated to meet changing needs and loss experience. Almost 25,000 certificates were in force under all group insurance programs in 1974. Coverage was available to members of student chapters, and special plans were provided for members over 65 years of age.

In 1957, ASCE considered a special program covering "errors and omissions insurance," for members in private practice, but the idea was dropped. In 1974, ASCE was exploring other group service such as supplemental pension annuities, automobile and homeowners insurance, and dental insurance.

Joint Economic Welfare Efforts

In 1973, ASCE considered joint action with other societies regarding the economic welfare of the civil engineer. A set of "Guidelines to Professional Employment for Engineers and Scientists" had been promulgated through collaboration between EJC and the NSPE. The recommendations covered recruitment practices, terms of employment, professional development, and conditions of termination and transfer. ASCE, the American Institute of Chemical Engineers, ASME, IEEE, and NSPE promptly endorsed the guidelines, and, as of mid-1974 a total of 25 organizations had followed suit.

Leadership by the American Chemical Society early in the 1970s had produced an ambitious "portable" pension plan proposal, which ASCE and other engineering societies immediately supported. This was under review by the Internal Revenue Service in 1973 and, if found feasible, would require some years to implement. In the meantime, ASCE initiated a Joint Pension Committee as a mechanism for cooperative effort toward improving "the legislative and regulatory climate in which engineers/scientists pension plans exist."

Although unemployment was not a serious problem to civil engineers in 1974, it was sufficiently prevalent in other engineering sectors to merit attention. EJC proposed that the U.S. Department of Labor establish a cooperative engineering job placement service under professional control, and ASCE supported the idea. The plan, if adopted, would operate with government support rather than in competition with employment services offered by government agencies.

Young Engineers

The founders of ASCE were mature, prestigious engineers, established in the top rank of their profession. Because they were preoccupied by organizational problems, they did not concern themselves with the upcoming generations until 1873, when they established the Junior grade of membership. The founders understood the term "Junior" as not referring to the age of a person but to classification in the Society, and they defined it as those engineers whose "professional experience has had a more limited scope than theirs."

Still, the Junior grade was a breakthrough, even without the voting privilege. Requirements were two years of engineering practice, or a college degree and one year of practice. About 5% of the members were Juniors in 1875; 10 years later the number had risen to 11%. In the 1870s, the Society also discussed the idea of student members but took no action.

At the 1887 Annual Convention, Robert E. McMath proposed a membership grade for students. The Board appointed a committee to consider and report on "the advisability of adding a new grade of membership to be called 'students.'" The committee recommended creating a Student grade for those under 18 years of age with a minimum of one year of study in a technical school or two years of engineering study and practice, with advancement to higher grade required in seven years.

Most of the 1888 Annual Meeting was devoted to discussion of this proposal. Those opposing it claimed that

> this Society is not an institution for primary education, but is intended to be an association of skilled and experienced engineers for mutual improvement by the interchange of ideas and experiences of a character more advanced than can be fully understood by novices in professional work. Admission to any grade of membership should be contingent on experience had and work accomplished, and not on the mere desire to learn and the hope of future benefit by association.

This reasoning prevailed. Although the Society was later to authorize student chapters, there was never a student grade of membership up to 1974.

Francis Collingwood (Secretary, 1891–94) in 1894 presented a $1,000 bond to the Society to fund the Collingwood Award for juniors. In his letter transmitting the gift he said:

> The most rapid advancement of the Society will be promoted by interesting the young members of the profession in its affairs. The professional advantage accruing from friendly intercourse, and ... from well-written papers presented at our meetings, is not ... appreciated by our older members and it is not strange that there (is a tendency) among juniors to lose interest in the Society.

The award failed to attract high caliber papers at first, and Collingwood urged in 1899 that ASCE hold meetings of juniors of the Society for reading and discussing papers. Members sharing his concern for young engineers were very much in the minority well into the twentieth century.

Student Chapters

At the close of World War I, a special Committee on Student Branches recommended forming student chapters. M.E. Cooley, Dean of Engineering at the University of Michigan, filed his written dissent. Dean Cooley saw the student chapter as narrowing the student's viewpoint. He urged instead "the breadth that comes from the study of other than technical subjects" in order to produce engineers with "vision-generals, if you like—qualified to coordinate and direct the great special forces of individuals." Nevertheless, the Board adopted in January 1919 the recommended "Regulations for Student Chapters," and the move turned out to be a wise one.

Beginning in 1921, Student Chapters appeared at Stanford University, the University of Cincinnati, Rensselaer Polytechnic Institute, Drexel University, Iowa State University, Pennsylvania State University, the University of Pennsylvania, and Washington University–Saint Louis. By 1925, there were 71 chapters, with a total enrollment of 4,107 members, Table 5-3 shows the remarkable effect on membership in the entrance grade of the Society. Chapters proved to be vitally important in the transition from academic training to the professional practice of civil engineering.

The Society authorized Student Chapters only at institutions in which the civil engineering program was approved by the Board of Direction, and accreditation by the Engineers' Council for Professional Development (ECPD) was always accepted as adequate qualification. Where the curriculum was not accredited by ECPD, the Society authorized student clubs, and nine

Table 5-3. *Effect of Establishing ASCE Student Chapters on Attracting Entrance-Grade Members*

		Entrance-Grade Members	
Year	*Student Chapters*	*Number*	*Percentage of Total*
1875	0	73	4.8
1920	8	506	5.4
1925	71	771	6.8
1930	96	2,329	16.4
1935	110	3,046	20.4
1950	128	9,667	38.1
1974	183	25,530	37.3

were in operation in 1974. Total membership in all chapters and clubs at that time was somewhat in excess of 10,000.

A 1956 study showed that ASCE's Student Chapter plan compared favorably with the student member systems of the other four Founder Societies. At that time, almost nine out of 10 of the Junior Members entered the Society when they graduated.

Although only about half the Student Chapter members became members of the Society at graduation, many applied for admission later. The vitality of the chapters was directly related to the enthusiasm and leadership of the faculty adviser. In some cases, all graduating members applied for membership. Student Chapter policies and operations have been administered by a Committee on Student Chapters since 1923.

Younger Member Groups

As the number of young engineers grew, they gained ASCE recognition. First designated as a Junior, in 1873, the grade was changed to Junior Member in 1949, and to Associate Member in 1959. Voting status came in 1947, and eligibility for national office three years later. A 1947 *Civil Engineering* editorial on the constitutional amendment referendum on the vote as well as a dues increase stated in part:

> Here is the younger members' frankly stated view that it is not enough for the Society merely to provide good tools, in the form of outstanding technical training and engineering knowledge, but that they also expect it to strive jointly for a proper environment in which they can utilize those tools advantageously—an environment of professional recognition.

When the ballots were counted, the Juniors had the vote, but the increase in dues did not receive the necessary majority.

In 1935, when Juniors made up about 20% of the total membership, the Board of Direction decreed that in so far as possible a Junior be given a place in every Society committee, including the Local Sections and Technical Divisions, so committees could benefit from their point of view.

Another policy adopted in 1951 requested every Local Section to appoint at least one Junior Member to each of its committees, and to delegate complete responsibility to a Junior Member for at least one meeting each year. Other actions provided for young members to be appointed as junior Contact Members for Student Chapters, Local Section Conference delegates, and as members of other committees. A Committee on Juniors appointed in 1930 became the Committee on Junior Members in 1950 and the Committee on Younger Members in 1959.

Table 5-4. *Entrance-Grade Participation in ASCE Activities in 1956 and 1970 (percent)*

Measure of Participation	*1956*	*1970*
Entrance-Grade Percentage of Total Membership	45	42
Participation by Entrance-Grade Members		
Authorship—journals	10	38
Authorship—*Civil Engineering*	13	26
Technical committees	7	16
Local Section committees	29	—
Local Section officers	24	35

Several surveys have been done to determine the actual involvement of young members in activities of the Society. Table 5-4 is a comparison of data gathered in 1956 and 1970:

In 1970, the Board of Direction requested annual reports on young members' participation in Society affairs, even to the extent of setting age 35 years as the upper limit for a "Younger Member." Data for 1971 and 1972 appear in Table 5-5.

Forums

Thirty-one years after Francis Collingwood urged meetings of Juniors of the Society for the reading and discussion of papers, Juniors in the Los Angeles Section, inspired by A.W. Dennis, formed the first Junior Forum. They met for an hour preceding each meeting of the Section. In addition to presenting papers, the Forum gave the Juniors and Student Chapter members opportunity to experience the professional affairs of the Society.

The idea spread. Following the 1930 Los Angeles Section Junior Forum, similar organizations began in the San Francisco Section (1932), the Metropolitan Section (1933), the Illinois Section (1937), and the Sacramento Section (1939). By 1950, there were 10 Local Section subsidiaries for Juniors, and in 1973 there were 14, all designated as Associate Member Forums. Junior Branches in the Central Illinois, Cleveland, Illinois, and Northwestern Sections did not survive.

The Junior and, later, Associate Member Forums were generally effective. In addition to technical seminars, the forums offered career guidance; continuing education (including review courses for candidates for registration examination); and manpower programs to encourage disadvantaged people in technical careers. The Forums also drew attention to economic and employment problems in the profession. The Forums initiated several important actions at the Local Section and even national levels.

The Forums in the Los Angeles, Sacramento, San Diego, and San Francisco Sections were particularly energetic. These four Forums achieved some

Table 5-5. *ASCE Younger Member Participation in 1971 and 1972*

Measure of Participation	*1971*	*1972*
Members and Associate Members under 35	32	33
Paying Local Section dues	45	46
Committee participation		
National technical	7	12
National professional	13	17
Local Section participation		
Section officers	33	38
Branch officers	35	59
Technical groups	47	57
Associate Member Forums	73	99

unity through annual regional joint conferences. In 1968, the conference theme was "The Professional Engineer and the Union," and that meeting resulted in a Los Angeles Section's committee report that called for national and regional guidelines for engineering fees, salaries, and employment conditions.

The 1969 Regional Associate Member Forum Conference had a "Unionism and Professional" theme, and it generated strong support of the guideline proposal in California and elsewhere across the nation. This membership demand was definitely a factor in the 1970 action at the national level of ASCE emphasizing salary and employment problems.

But the California Associate Member Forum campaign had an even greater impact. The four Forums combined to elect a Director to represent District 11 in the Board of Direction who was knowledgeable about the unionism problem and aware of the need for prompt action. Their candidate was nominated but lost to the official nominee by only 127 votes in almost 4,000 ballots cast. The incident drew attention to certain shortcomings in the nomination and election procedures, which prompted the Board of Direction to explore reform measures.

Many members think holding younger members apart from their elders complicates the progression of the young into the mainstream Society leadership. However, the forums apparently interest and challenge younger members, and many ASCE leaders found their first opportunities to serve the Society through these groups.

Women in the Society

The first woman to become a member of the American Society of Civil Engineers was Nora Stanton Blatch De Forrest Barney, said to be the first woman in the United States to receive a civil engineering degree (Cornell University, cum laude, 1905). Nora Blatch—the granddaughter of Elizabeth Cady Stanton, a pioneer in the women's suffrage crusade—became a Junior of the Society in 1906. She worked in the New York City Engineering Department and contributed a technical discussion to the 1906 volume of *Transactions.* In 1909, she married Dr. Lee De Forrest, inventor of the vacuum tube. When she tried to advance to the corporate grade of Associate Member in 1915, she was denied. She then petitioned for a peremptory mandamus that would have required the Society to grant her admission to the Associate Member grade. The petition was denied, and Blatch was dropped from membership for failure to qualify for advancement to corporate status in the required time. She later became an architect. Eleven years later, in 1926, Laura Austin Munson was elected to the grade of Junior. The occasion was marked by a special item in *Proceedings*:

> The question of opening Society membership to women has received considerable attention in the past and opinion has been divided. Some have argued the question from the standpoint of expediency and some from justice, but none from exclusiveness. ... Frequently members took the view that women might be engineers—in fact, that some already were engineers but that the number was small from the physical limitations usually required of one in the profession. Thus, in the past, the discouragement of women as members of the Society was considered to avoid inconvenience and embarrassment to all concerned.
>
> This action by the Board is the result of the usual complete scrutiny given all applicants and is significant of the trend of the day. ... Women [are] accorded full equality with men in professional as well as public life.

Elsie Eaves, admitted as a Junior in 1926, was the first woman to progress through the grades of Associate Member, Member, and Fellow. In 1962, when she achieved Life Member status, she was the manager of the business news department of McGraw–Hill Publishing Company.

A 1949 *Civil Engineering* item noted that about 50 women had been admitted to membership up to that time, and that 43 were then enrolled. Most of these had joined since 1940, possibly as a result of the emphasis on technology during World War II. The majority of women members were employed in federal and state agencies, about a dozen in private and industrial practice, and two in teaching.

Although member gender is not noted in ASCE records, at least 226 women could be identified in the roster as of March 1974, about 0.3% of the total membership. In the 1970s, there was a noticeable increase in the enrollment of women in civil engineering programs, and it was not uncommon to find young women serving as officers of Student Chapters. However, women were rarely prominent in Society activities at the Local Section and national levels.

Professional Practice Problems

A professional society must identify situations that might interfere with the effective performance of its practitioners, and it must seek solutions to those problems. ASCE's Code of Ethics is one means to accomplish this. Some problems are managed through administrative procedures, as in the Committee on Professional Practice, where equal representation ensures recognition of the interests of engineers in both public and private practice.

A few examples show how ASCE has coped over the years with a wide range of incidents. Early in 1923, Secretary of the Interior Hubert Work asked Arthur P. Davis, Past-President, to resign his post as Director of the U.S. Reclamation Service. Secretary Work justified his replacement of Davis with a former Governor of Utah, a nonengineer, on the grounds that the engineering phases of the Service were superseded in importance by the business and economic aspects. The summary manner in which Davis was dismissed, together with the implication that engineers were not capable of managing important business functions, elicited a strong reaction from ASCE.

The Board of Direction formed a special committee to investigate the case, and it sent a strong letter of protest to Secretary Work. The letter was circulated widely, and the Society urged Local Sections to express their displeasure to their legislators. The National Civil Service Reform League entered the fray on the side of ASCE, emphasizing that Secretary Work had abolished the Civil Service position held by Davis, creating a new position outside the Service with the same responsibility under a new title.

The storm grew to such proportions that Secretary Work set up a special Fact-Finding Commission "to investigate the whole system of Government methods in reclaiming arid and semi-arid lands by irrigation." By this gambit he allayed much of the criticism, and in the spring of 1924 he appointed Elwood Mead, an engineer and member of ASCE, to succeed the ex-governor as Commissioner of Reclamation. The Society commended Work for this action.

The magnitude of the reaction to Davis's dismissal may have reflected his stature as a Past-President of the Society. This was not a factor in 1935, however, when the Board again took note of unfair treatment of several members employed in public agencies who were deprived of their jobs for doubtful cause.

In January 1935, the Board of Direction approved the principle that the Society should defend its members in such cases. In April of that year, ASCE adopted detailed conditions and procedures for defending members "against unjust accusation or dismissal, or oppression, without justification or proper hearing." The procedure involved the Local Sections, a personal investigation by a Vice President of the Society, with delegation of powers of adjudication and action to the Executive Committee of the Society.

ASCE applied the procedure almost immediately. The federal Public Works Administration (PWA) asked a state director to resign and declared a former associate in consulting practice ineligible for participation in PWA projects. The Society obtained a rehearing and proved the actions against both engineers founded on false information. The Administrator of PWA rescinded the punitive rulings.

In 1936, ASCE also defended the interests of members employed by state agencies in Pennsylvania and Delaware. After that, both the policy and the implementation process appear to have been lost, only to be unearthed in the course of this historical research. Neither of the two executive secretaries who served through the period 1945–72 was aware that the policy existed, and there were several instances during those years in which it might have been exercised. The policy was restored to official notice in 1973.

The Society has also acknowledged responsibility of the profession for actions of its members that may have been contrary to the public interest. In 1923, the government filed a number of suits, with members of the Society among those indicted, for alleged fraud in connection with certain contracts for the construction of World War I National Army Cantonments. The Board of Direction of the Society urged the President, Congress, and the Attorney General to bring these charges immediately to trial "in order that the guilty be punished and that the innocent may be freed of the serious accusations which have been made against them."

ASCE has consistently advocated the principle of Civil Service. In 1935, for example, there was cause for concern about possible erosion of Civil Service standards and procedures. The Society's Field Secretary in Washington conveyed specific recommendations in this matter to the proper authorities. A year later, ASCE formally endorsed the merit system and adopted the U.S. Civil Service classification of positions for use in future Society studies of civil engineering employment and compensation.

The projects of consulting engineers keep them in the public eye, and at times subject to unfair criticism. As early as 1914, an overzealous Grand Jury impugned the integrity of three reputable consultants to the New York City Board of Water Supply. The Society issued a sharp rebuttal and condemned the Grand Jury action, sending the statement was to the authorities involved and to the press. Civil engineers in public service are sometimes victimized by political campaign fund solicitation. In 1926, the Society condemned this practice, and termed such solicitation of funds or endorsement thereof on the part of a state department head to be "wrong and reprehensible."

From time to time, engineers get unfavorable press because of questionable practices in the promotion of professional services by consulting engineers. In 1956, commission agents offered to secure engagements for consulting firms. ASCE made a terse and timely pronouncement saying this action was in conflict with the Code of Ethics provision declaring it unethical "to use influence or offer commissions or otherwise to solicit professional work improperly, directly or indirectly." This action resolved the problem.

In 1970, the problem of campaign funds reached public attention when a Nassau County, New York, newspaper reported that some contributions might be associated with the award of professional service contracts to architects and engineers. Under ASCE leadership, architectural and other engineering societies promptly developed a set of guidelines that clarified the issue beyond question:

> An engineer who makes a direct or indirect contribution in any form under circumstances related to his selection for professional work shall be (a) subject to disciplinary action by the Society, and if appropriate, (b) reported to the public authorities.

ASCE's Board adopted a policy statement embodying this admonition in 1971, providing guidance for consultants and political campaign fund solicitors.

The profession suffered a loss of public esteem in 1973 when a federal investigation of corruption in Maryland implicated three members of the Society in the irregularities resulting in the resignation of U.S. Vice President Spiro T. Agnew. The investigation focused, in part, on political contributions that allegedly had been made in return for the award of contracts for engineering services.

The three members were expelled from the Society. In addition, a Task Committee on Professional Civic Involvement condemned immunity concessions to transgressors who implicate others, affirmed the 1971 guidelines, endorsed legislation requiring full disclosure of all campaign contributions, and supported state registration boards in suspension or revocation of licenses as punishment for professional misconduct.

The very nature of public works projects ensures continuing professional practice challenges for the civil engineer. ASCE is willing and able to serve as a medium for the prevention and control of such obstacles.

Honors, Awards, and Fellowships

All professionals cherish peer recognition of competence and noteworthy accomplishment, and ASCE award programs meet this need more than adequately. ASCE's highest honor, Honorary Membership, has had essentially the same requirements since 1874. This honor "acknowledges eminence in some branch of engineering" with at least 30 years of experience; in 1974, the wording was changed to "acknowledged eminence in some branch of engineering or in the arts and sciences related thereto, including the fields of engineering education and construction." Since the first five Honorary Members were elected on March 2, 1853, only 259 have been so designated through the next 121 years up to 1974. Among the engineers so honored are Herbert Hoover, Field Marshall Ferdinand Foch of France, Baron Christian Phillipp von Weber of Germany, Sir William Henry White of the United Kingdom, Karl Imhoff of Germany, Baron Koi Furuichi of Japan, and Professor Luigi Luiggi of Italy.

In 1872, George H. Norman contributed about $1,200 to fund an annual award of a gold medal for "the best essay on engineering subjects." Award rules were drafted and dies prepared for the medal according to Norman's design. The first award was made in 1874 to J. James R. Croes for his paper, "Memoir of the Construction of a Masonry Dam."

The 1881 Annual Report of the Board of Direction noted that the Norman Medal was the only Society prize, and it invited gifts to fund additional awards. Thomas Fitch Rowland responded immediately by endowing, in 1882, the prize that bears his name for papers on engineering construction. This was followed, in 1894, by the Collingwood Prize.

In 1912, the Society established the J. James R. Croes Medal and the James Laurie Prize, and in 1924, the Rudolph Hering Medal, endowed by the Sanitary Engineering Division. That same year, Past-President John R. Freeman funded the first ASCE Fellowship to encourage young engineers in research.

By 1960, 21 prizes, medals, and fellowships were awarded by ASCE. Efforts in the next several years to encourage additional awards were successful, and by 1974 the list had grown to 39, with reserves for these totaling almost $600,000. Most of the new awards initiated in the 1960s were endowed through the Technical Divisions.

ASCE cooperates in four important awards made jointly with other societies. These include the John Fritz Medal, instituted in 1902; the Washington Award, 1916; the Hoover Medal, 1929; and the Alfred R. Noble Prize, 1929. Four of the Founder Societies are involved with all these awards, the Western Society of Engineers with the Alfred R. Noble Prize and the Washington Award, and NSPE with the Washington Award.

The *Official Register* includes the names of all award-winners each year.

During its first 80 years, ASCE's efforts to create a professional environment for its members were both impromptu and disconnected. After 1930, the era of the Great Depression, there was almost too much planning of programs that the Society was not fully equipped to administer. After years of experience and unprecedented commitment of funds and staff, a productive future is assured.

CHAPTER 6

Advancement of the Civil Engineering Art

The 1852 Constitution of the American Society of Civil Engineers and Architects states the Society's objectives:

> The professional improvement of its members, the encouragement of social intercourse among men of practical science, the advancement of Engineering in its several branches and of Architecture, and the establishment of a central point of reference and union for its members.
>
> Among the means to be employed for attaining these ends, shall be periodical meetings for the reading of professional papers, and the discussion of scientific subjects; the foundation of a library, the collection of Maps, Drawings and Models, and the publication of such parts of the proceedings as may be deemed expedient.

Most members felt technical matters should be the sole concern of the Society. Such emphasis is not surprising, in view of the status of applied science in 1852. Over time, there has been a shift to emphasis on professional development.

ASCE recognizes significant contributions of the members by many honors and awards, and it strives to offer meetings that attract the greatest possible attendance and member participation. Society publications are renowned worldwide. The specialized Technical Divisions have fostered hundreds of committees to address specific technical problems. The Engineering Societies Library has provided a central point of reference for the entire engineering profession, not just civil engineering.

We Stand on Their Shoulders

When ASCE became an official society in 1852, the Canal Era was declining, and a memorable group of civil engineering pioneers had already passed from the scene. Such great engineers as the Loammi Baldwins, father and son, Canvass White, Benjamin Latrobe Sr., William Roberts, Lewis Wernwag, James Finley, James Geddes, William Howe, and Benjamin Wright were gone. Benjamin Wright had been a participant in the 1838 effort to form a national society, but he died 10 years prior to the 1852 movement.

Three Honorary Members of ASCE—John B. Jervis, Moncure Robinson, and Stephen H. Long—were leaders in the transition from the canal to the railroads, as were Past-Presidents James Laurie, Horatio Allen, W.J. McAlpine, W. Milnor Roberts, and Octave Chanute. Other members of the Society who contributed to the early development of the railroad included Joseph G. Totten, Theodore M. Judah, and Albert B. Rogers.

Among the famous pioneer bridge-builders were Honorary Member Squire Whipple and Past-Presidents Albert Fink, Julius W. Adams, Elmer L. Corthell, George S. Morison, Henry Flad, and Thomas C. Clarke. Many innovations in bridge design originated with ASCE members. Jacob H. Linville and L.G. Bouscaren were the first to use wrought-iron members. Other contributions were John A. Roebling's steel-wire suspension bridge; William Sooy Smith's pneumatic pile and James B. Eads's pneumatic pier; the arch bridges of Lefferts L. Buck, and the cantilever designs of Charles Shaler Smith and Charles C. Schneider. Eads earned fame for both his Mississippi River Bridge at St. Louis and his daring execution of the South Pass jetties in the Mississippi, which made New Orleans a major port.

Sanitary engineering was another area of innovation by the first leaders of ASCE. Past-Presidents James P. Kirkwood, Alfred W. Craven, Julius W. Adams, Rudolph Hering, and Ellis S. Chesbrough were the forerunners of a distinguished cadre of innovators in water supply, sewage works, and public health technology. Kirkwood introduced the art of water filtration to America, Adams and Chesbrough were the first designers of public sewer systems for Brooklyn and Chicago, respectively, and ASCE was organized in the office of Alfred W. Craven, chief engineer of New York's Croton Aqueduct. Chesbrough was also experienced in water supply, and he may have been the first city engineer in serving Chicago in that capacity. Charles Payne in Providence and John P. Davis in Boston also provided the function of the city engineer, though not under that title.

Just as the canal builders transferred their talents to the rising needs of the railroad industry, so did they move on to face the challenges posed by urban growth in the late nineteenth century. Laurie, Adams, Kirkwood, Craven, and Chanute typified this generalist view of the early civil engineer. The demands of the burgeoning America were so diverse that specialization was a luxury.

From the days of Hadrian in ancient Rome, intense public works have been plagued by political chicanery, exploitation of public rights and property, and fraudulent schemes by some promoters of commercial and industrial enterprises. This was the case during both the Canal Era and the Railroad Era in America.

Many of the early leading members of ASCE worked on projects conceived by quick-profit entrepreneurs, and instant conflict was often the result. At the 1873 Annual Meeting of ASCE, Martin Coryell pointed to the mismanagement of railroads by political manipulators and Wall Street speculators who had no interest in the scientific management of transportation. John B. Jervis, Mendes Cohen, Alfred F. Sears, and William R. Hutton were also among the ASCE members who exposed the incompetence and corruption in railroad management. Cohen and Hutton resigned their top-level administrative positions rather than remain privy to practices that were unprofessional or worse. This spirit of integrity and rugged independence extended into the public sector as well, as exemplified in 1860 by the public resistance of Alfred W. Craven to interference by the mayor of New York in the engineering aspects of the development of the city's water supply.

The inept railroad and industrial management prior to the Civil War created a vacuum filled by a cadre of civil engineers who adapted their education and training to management. Engineers knew the importance of sound basic data, and they could compile statistical information so it could be analyzed, interpreted, and applied to management. Professional management was actually "invented" by the civil engineers of this era.

In 1861, John B. Jervis published his "Treatise on Construction and Management of Railways." During the Civil War, Montgomery Meigs and Samuel Felton employed the science of logistics to allocate and move men, materials, and weaponry.

Albert Fink was foremost in this group. His contributions to the statistical and economic analysis of railroad construction and operation far transcended the early bridge truss design for which he is best known. As vice president of the Louisville and Nashville Railroad, in 1873 he published a report, "Cost of Transportation," in which he developed several of the theories and indices fundamental to transportation economics.

Octave Chanute demonstrated the efficacy of the engineering approach in his 1873 rescue of the Erie Railroad from financial disaster. As general manager, Chanute successfully reorganized and rebuilt the line. Arthur M. Wellington was also eminent in this field. His 1877 book, *The Economic Theory of the Location of Railways*, became a standard reference on the subject. Ashbel Welch, another accomplished railroad executive and engineer, made significant contributions to railroad safety.

By 1880, the role of the civil engineer as a professional executive and administrator was well established. A number of ASCE members were railroad presidents, among them Ashbel Welch, Mendes Cohen, Samuel Felton, Sidney Dillon, Eckley B. Coxe, and Alexander J. Cassatt. Many more distinguished themselves in such top-level managerial positions.

The engineer–manager also found opportunities in other areas of industrial enterprise. Alexander L. Holley, James B. Eads, and Martin Coryell became widely known as such consultants. In the public sector, James B. Francis, Alfred Noble, Frederick P. Stearns, and others filled responsible administrative posts in various government agencies.

Some highly respected civil engineers who practiced between 1850 and 1880 chose not to become affiliated with ASCE, but the number was small. Several were outstanding, including Herman Haupt, J. Edgar Thompson, Benjamin H. Latrobe Jr., John A. Devereux, George Stark, and Charles Ellet. These men had the same dedication to duty, professional conviction, and initiative that characterized the top civil engineers of the time.

American engineering art was nurtured by a relatively small group of individuals who sought better ways to provide the public works and services sorely needed by a young nation. Any organization that brought these engineers together to exchange ideas and to evaluate innovative proposals was beneficial. In providing the venue, ASCE became a national influence in the building of America.

The Society's efforts were set in a complex system. The system's principal elements were the Society's many kinds of meetings, wide range of publications, multitude of Technical Committees and Councils, growing list of specialty fields, and several major administrative committees (e.g., Technical Activities, National Meetings Policy and Practice, and Publications).

Meetings, Conventions, and Conferences

During the struggling pre–Civil War days of the Society, its only real activity was "the encouragement of social intercourse among men of practical science." There were 19 meetings held in the years 1852–55. All took place in the office of Alfred W. Craven, chief engineer of the Croton Aqueduct Department, Rotunda Park, New York City. Attendance was woefully small, averaging only six for the eight meetings held in 1853 and even less for the six held in 1854.

President Laurie's description of an elevated railway plan for "The Relief of Broadway," in 1853, was the first technical program. Similar presentations covered a range of subjects. In March 1855, William H. Talcott read the first formal paper placed in the files of the Society: "Results of Some Experiments on the Strength of Cast Iron." This was the last meeting to be held until the Society was "resuscitated" in 1867.

The revitalization meeting on October 2, 1867, was in the office of C.W. Copeland at 171 Broadway, and adjourned to a session at 76 John Street on October 9. On November 6, 1867, the Society met for the first time in its own headquarters in the Chamber of Commerce Building at 63 William Street. The next meeting, on December 4, 1867, was noteworthy for several reasons: ASCE elected 54 new members, it adopted a semimonthly meeting schedule, and James P. Kirkwood gave his "Address of the President," the first paper published by the Society. Occasionally, nonresident members attended, but the gatherings were still quite local. The following notice emphasized "the encouragement of social intercourse among men of practical sciences":

> May 10, 1869
>
> Dear Sir:
>
> Our Society has of late been impressed with the importance of annually gathering its members in Convention for the interchange of ideas and the discussion of professional subjects, as well as those relating to the extension of its own field of usefulness.
>
> These objects cannot be attained by means of the twenty three regular meetings, held during the year, on account of the distant residence of many of its members, but they require the appointment of some day on which all can meet and strengthen that bond of sympathy which should animate hearts engaged in the furtherance of a common cause.
>
> It is proposed, therefore, to hold the first of such a series of conventions on the 16th day of June, 1869, commencing at 10 a.m.
>
> Through the courtesy of the Chamber of Commerce, the meeting will be held in their Assembly Room. It will occupy most of the day, and will be followed by a suitable collation.
>
> The Society relies on the individual efforts of its members for sustaining the interest of the occasion, and desires that you will either come prepared to read a brief professional paper or in case of unavoidable absence will forward such to the Secretary—or else be prepared to make some statement of engineering experience calculated to prove of general interest.

Fifty-five members registered for this first of many ASCE conventions. Honorary Members John B. Jervis and Squire Whipple gave informal statements, others presented five formal papers, and the Convention closed with a dinner. This pattern continued for many years.

When the second Annual Convention was not as successful as its predecessor, the Society decided to hold the annual conventions in other parts of the country, to emphasize its national character. After a successful experiment in Chicago in 1872, the fifth Annual Convention in Louisville drew 79 Members and Fellows, the largest number of civil engineers ever assembled in North America up to that time.

The Society adopted the Annual Convention in 1869 as a supplement to the Annual Business Meeting in January and the regular monthly meetings at Society headquarters. In 1922, the Society adopted a schedule of three conventions per year instead of one. That same year, ASCE authorized specialized areas of civil engineering (e.g., sanitary, structural, highway, construction) as separate Technical Divisions, and the two actions were no doubt related. In

1929, the regular monthly meetings ceased and the schedule became the Annual Meeting in January, with the spring, summer, and fall conventions.

In 1950, ASCE changed the frequency of its national gatherings. The Annual Meeting was changed to October, and only two conventions were now held, in February and June. ASCE held no conventions during the World War II years of 1944–45, and it held between one and three conventions between 1932 and 1961.

Until 1930, the programs of the meetings were limited geographically. At the Cleveland Convention that year, a regional committee handled the program and general management, with the business management left to the host Local Section. By this time, the Technical Divisions were the primary source of program material. This pattern continued for about 30 years.

In the late 1950s, when the three national conventions were annually attracting a registration of 4,000 to 5,000, a new trend developed. The Technical Divisions produced periodic journals for its technical papers and discussions, which gave them a certain autonomy. The national meetings produced about 450 to 500 papers per year, but the capacity of the Divisions was much greater. To provide further outlet for their publications, the Divisions began to hold secondary conferences, sometimes alone and sometimes jointly with other Divisions or other engineering organizations. These Division conferences were usually, but not all, devoted to a specific problem area, such as electronic computation, weather modification, jet-age airports, flood plain regulation, shear strength of soils, pollution abatement, and economics. By 1960, the Divisions held six to eight conferences a year, drawing between 1,200 and 1,500 attendees annually.

In 1962, ASCE tried a new formula to emphasize national meetings. Instead of attempting to cover the interests of all Technical Divisions, each of the national meetings concentrated on one of the four areas of civil engineering: structural, water resources, transportation, and environmental. Broad-interest Divisions, such as the Construction and the Surveying and Mapping Divisions, could offer papers under any of the topics.

Although it was formally adopted by the Board of Direction, the plan was never actually implemented. Some Technical Divisions insisted on participation in every national meeting, regardless of the relevance of their sessions. Other Divisions held their own conferences in competition with national meetings. The Annual Meeting concept was also not implemented.

By the early 1970s, attendance totaled about 4,000 at the four annual topical meetings. Seven to 12 Technical Division conferences drew 1,500 to 2,500 attendees yearly, and the Divisions were also participating with other organizations in many joint specialty conferences. In 1973, the Committee on National Meetings Policy and Practice was appraising the situation, hoping to increase attendance and reduce the number of meetings.

Before 1957, the only exhibits displayed at ASCE meetings were noncommercial, mainly project models and photographs. That year, ASCE included a trial commercial exhibit featuring new developments in civil engineering equipment at the Annual Meeting in New York. The exhibit added value to the Convention and produced some income.

The 1957 Civil Engineering Show was modestly successful—sufficiently so to merit continuance at Annual Meetings. After a few years, however, the federal Internal Revenue Service alleged that the exhibit income was unrelated to the tax-exempt purposes of the Society, and it filed a claim for $47,978 in income taxes for the years 1969–70. The claim was successfully contested, and the complaint was withdrawn, but the incident raised questions. In 1965, the Board of Direction concluded that the values hardly justified the effort, and it discontinued commercial exhibits in conjunction with basic or specialty conferences, while permitting any educational or other noncommercial exhibits arranged by the local committee.

The interest in commercial exhibits revived a few years later, when the Water and Wastewater Manufacturers offered to manage an equipment exposition at the 1973 Annual Meeting. The Board amended its policy to sanction such functions at national meetings, when requested by the local committee. The 1973 show in New York City was only moderately successful. However, there was a slight surplus of income over expenses, and the Society decided to explore other approaches to exhibits as a source of income.

Many national meetings of the Society were noteworthy for one reason or another, but a few stand out. The 1876 American Centennial Exposition in Philadelphia provided an unique opportunity to direct public attention to the promising young engineering profession. The 1876 Annual Meeting was held in Philadelphia in conjunction with the exposition, directed by a special ASCE Centennial Commission and a Resident Secretary. A general exhibit featured 1,900 displays, including pictures, books, manuscripts, maps, drawings, and even full-scale models. The exhibit was the means of "bringing together, expeditiously and effectively, engineers from all parts of the world, placing them in communication with manufacturers and supplying them with letters of introduction and data concerning such engineering works as they were desirous of visiting." Many foreign engineering displays were also shown at the exposition, and the event was the first international recognition of American engineering.

A highlight of the program of this Annual Meeting was a paper by ASCE Vice President Theodore G. Ellis, in which he spoke of American engineering in the future and, specifically, the use of aluminum as an engineering material. In part, he said:

> With a metal only 2½ times as heavy as water, 6 times as strong as steel, as easily worked as iron, and almost indestructible in air or water, conceive what would be the possibilities of engineering skill, to say nothing of the revolution in all the tools and utensils of ordinary life. Bridges of a mile span would be entirely possible, railway trains for passengers could be reduced to at least one-fourth their weight, and the navigation of the air (all except the alighting, in

> which I should never have great confidence), would become practicable. With a motive power which would not be obliged to carry so great weight of fuel as our present steam engines require, speed of navigation might easily be doubled.

The 1876 Annual Meeting was so successful that the United States Centennial Commission gave the Society an award "for the very large and important exhibition, and for the great service rendered by the Society to the art and science of engineering."

At the Paris Exposition in 1878, ASCE received a Diploma of Honor for its exhibit. ASCE was responsible for the Civil Engineering Division program in the International Engineering Congress at the Columbian Exposition in Chicago in 1893.

ASCE considered an exhibit at the Paris Exposition in 1900, but instead the beginning of the new century was commemorated by holding the 1900 Annual Meeting in London in July. The Institution of Civil Engineers offered facilities and hospitality for the occasion, and 68 members attended.

At the request of the Louisiana Purchase Exposition, ASCE financed and conducted an International Engineering Congress in conjunction with its 1904 Annual Meeting. This event was a feature of the St. Louis World's Fair, attracting a registered attendance of 876. The program was truly international, with 51 papers from the United States, many others from France, England, Holland, and Japan, and single papers from six other European nations.

The 1924 Annual Convention in Pasadena, California, had a tragic aftermath. The train carrying a number of Society officers and members back East was in a serious collision and Secretary John H. Dunlap was so severely injured that he died a month later. The Society assumed a deep and lasting interest in the welfare of his wife and three sons.

In 1952, the 100th anniversary of the founding of ASCE was celebrated with great pomp and fanfare. However, the Centennial of Engineering, held in Chicago that September, was much more than a birthday celebration for ASCE. It was also a worldwide commemoration of the first century of engineering as a profession in America.

Major Lenox Lohr directed the Centennial with these aims: (1) highlighting the engineer's contributions in peace and war; (2) "personalizing" the engineer; (3) encouraging young people to pursue engineering careers; and (4) depicting the role of technology in the enhancement of the standard of living and national prosperity. The program included several hundred professional and technical papers and an exposition, and other engineering organizations held meetings in Chicago during the 10-day celebration, highlighting the profession. Total registration reached 27,964, including 2,600 ASCE members.

Former President Herbert Hoover, Hon.M.ASCE, gave the Centennial Day luncheon address, and ASCE President Carlton S. Proctor, representing the Society, received 66 gifts and greetings and congratulation from engineering societies throughout the world. As a tribute to the Society on its 100th

Figure 6-1. *ASCE Centennial Commemorative Stamp and First-Day Cover*

anniversary, the Post Office Department issued a special commemorative stamp during the Chicago Convocation. The stamp compared a typical 1852 wagon bridge with the George Washington Bridge in New York City, and it portrayed George Washington as a pioneer of civil engineering. An official engraved cover was provided for first-day mailing of the commemorative stamp (Figure 6-1).

Society Publications

When James P. Kirkwood assumed office as the second President of ASCE immediately following its rebirth in 1867, he lost no time in calling for services that would ensure the permanence of the Society. In his inaugural address he said:

> The nonresident members may many of them connect themselves with our Society from a kind of esprit de corps. ... But this link of union between the Society and those distant members who can but rarely enjoy ... its meetings ... or library ... must be maintained and nourished by some palpable food. The short papers to which I have alluded on subjects of professional interest or a selection from them, must be printed ... reaching and interesting the nonresident members.

Action to provide the "palpable food" was immediate. At that same meeting, on December 4, 1867, ASCE created a Publications Committee. (Supervision of publications was later combined with the duties of the Library Committee. However, there has been a standing Committee on Publications since

1891, except between 1923 and 1927, when there was a Committee on Technical Activities and Publications, and 1927 and 1930, when there was a Committee on Meetings and Publications.)

A few months later, the Society inaugurated a special fund for printing and disseminating papers. After considerable effort, 50 donors contributed $100 each in demonstration of "fellowship among the patrons and capitalists of public improvement and those interested in the onward progress of the sciences." In February 1870, the Society established the membership grade of Fellow to recognize all contributors of $100 or more. By 1882, the fund reached $10,000, and the Board determined that all receipts above that level would be transferred to the general revenues of the Society. By 1891, the Society accommodated publication expenses in the regular operating budget, and it abolished the Fellowship Fund.

From 1867 to 1873, technical papers accepted for publication were issued separately and irregularly as *Transactions* of the Society. A total of 57 papers was produced and distributed in this fashion. The following resolution, adopted early in 1873, began the system of handling technical papers still used a century later:

> *Resolved,* that hereafter every paper presented to the Society shall be immediately examined by the Library Committee, who shall decide whether it shall come before the Society. If yes, it shall be printed in cheap form and distributed to the members, with notice that discussion, written or oral, will be received, within definite limits as to time; at the expiration of which, said discussion, with the original paper, shall be referred to a special committee, with instructions to examine and recommend a final disposition of the same, with reference to the permanent proceedings of the Society.

A second action, also in 1873, provided for the Society's first regular publication. It was to comprise not less than 48 pages and to be issued on the second Wednesday of each month. The new monthly, *Transactions*, included all papers accepted by the Society together with all contributed discussions on them, as well as a section designated *Proceedings*, with notices and reports of meetings of the Society and the Board of Direction, news items, announcements, new book lists, membership actions, memoirs of deceased members, and library acquisitions.

The 1872 number of *Transactions*, Volume I, was issued in November 1873. Volumes I and II carried the "Transactions Papers" selected for publication during the period 1867–72. One was the 1867 Presidential Address of James P. Kirkwood.

The volume of material was more than sufficient to support the authorized monthly issues of 48 pages, and *Transactions* was produced with this frequency from 1873 to 1895.

From 1879 to 1892, a few extra copies of each paper were printed in advance and circulated to selected reviewers as a means of stimulating discussion. From 1892 to 1895, the Society published a *Bulletin* for all members. It carried all official notices and announcements, with abstracts of papers to be read at forthcoming meetings. Any interested member could request an advance copy of the complete paper.

January 1896 brought the first number of the new *Proceedings*, issued on the fourth Wednesday of each month except July and August. The *Proceedings* carried all items pertaining to Society business, meeting notices, minutes, and the like, and all papers accepted by the Publications Committee for subsequent meetings. The final drafts of these papers, with all discussion and correspondence, were then published in an annual volume of *Transactions* for the permanent record. For foreign members and others unable to attend meetings regularly, the Society published a newsletter supplement from 1924 to 1930. Content was limited to notices, news items, and short feature articles of general interest.

Except for the frequency of issue of *Transactions*, this publication program continued on a complimentary basis to all membership grades for the next 50 years. The growth in volume of published material justified semiannual publication of *Transactions* in 1887, and a quarterly schedule in 1909.

A major development was the introduction in October 1930 of the monthly magazine *Civil Engineering*, which provided valuable internal communication to the fast-growing Society (which then had 14,000 members). According to a notice in the *Annual Year Book*:

> It carries the more animated and graphic articles submitted for publication, and many Society announcements of official and semiofficial character which formerly were contained in *Proceedings*. The *Proceedings* were thus simplified, making more readily accessible those contributions to engineering literature of the more studious nature. The change contributes largely to a wider spread of knowledge of both Society and technical matters, providing, with *Proceedings*, two mediums different in style of expression and character of content.

Civil Engineering met its editorial objectives successfully without charge to members, its costs being offset by advertising revenue sufficient to sustain it through the Great Depression. As "The Magazine of Engineered Construction," its advertising income by 1955 not only covered all production and staff expenses but also contributed to the Society's general revenue. In 1972, the Internal Revenue Service claimed that surplus advertising and subscription receipts were taxable "as income unrelated to the exempt purposes" of the Society. The Society successfully resisted this claim. In 1970, the subtitle of *Civil Engineering* was changed to "The Magazine of Environmental Design and Engineered Construction," recognizing the trend toward environmental issues.

After more than 30 years using the name *Civil Engineering*, the Society discovered that a commercial periodical in England held prior claim to that title under copyright. The matter was promptly settled by a change to *Civil Engineering—ASCE,* with other nominal concessions. Oddly, not a single one of the then-70,000 readers commented on the revision.

Rising printing and mailing costs in the 1940s created economic problems with *Transactions* and *Proceedings*, both of which were almost completely subsidized from general operating revenues. The only income derived from these publications came from nonmember subscriptions. In 1947, ASCE imposed a modest charge for *Transactions* to reduce the subsidy.

By 1950, the amount of technical material being provided by the Technical Divisions for publication in *Proceedings* was accelerating. To reduce waste circulation, *Proceedings* no longer published on a monthly schedule, and beginning in February 1950 each technical paper or report was published as a *Proceedings—Separate. Civil Engineering* listed abstracts with an order form. Up to 25 *Separates* per year were furnished without charge to members, and additional copies cost 25 cents each. In 1954, this member service was modified. The Society allowed members to enroll in one Technical Division and to receive automatically, and without charge, all the papers sponsored by that Division.

The system was not popular. The proliferation of technical specialty organizations throughout the engineering profession is often a result of a one-time need for a suitable publication to serve a specialist field. A major adjustment was in order.

The Society resolved the situation by producing *Proceedings* in the form of a separate periodical for Technical Divisions, effective in January 1956. These *Proceedings—Journals* assembled the papers of each Division under its own review, and every member could receive any two journals without charge. Additional journals and separate papers were available for purchase. The action satisfied the Technical Divisions, which now had essentially all of the services and autonomy of an independent organization.

The *Proceedings—Journals* were still fulfilling their purpose in 1974, although economic considerations by 1966 required that they be issued on a subscription basis. The Air Transport, Highway, and Pipeline Divisions decided in 1969 to combine their separate Division journals into a single joint quarterly *Transportation Engineering* journal. The publication frequency of the journals varies from annual to monthly. Several journals—including those of the Engineering Mechanics, Environmental Engineering, Hydraulics, Geotechnical Engineering, and Structural Divisions—have acquired worldwide eminence in their fields.

When the *Proceedings—Journals* were adopted in 1956, *ASCE* provided a *Journal of the Board of Direction* that included professional material of broad membership interest. Renamed the *Journal of Professional Practice* in 1958, the publication found limited demand even though a complimentary mailing of an issue was made to all members in an effort to promote circulation. In 1972, its name was changed to *Engineering Issues—Journal of Professional Activities*, and a staff editor was assigned. The publication was making progress in 1974.

Since the 1950s, members and staff have paid attention to the appearance of the Society's publications. In 1953, ASCE adopted photo-offset printing methods for its technical publications and became the first technical society to use typewriter composition. The appearance of the text was less attractive than that produced by the previous hot-type methods.

By the late 1960s, computer-assisted photocomposition had become a reality, and ASCE used it to produce its 1968 *Directory* and 1971 *Official Register.* In 1972, computer typesetting for the journals improved their quality and stabilized costs. If these advances involved extra initial expenditures, they became economically supportable very quickly.

Until 1956, all technical papers published in *Proceedings* were reproduced as finally revised, with all discussion, in the *Transactions.* With over 250 papers, double publication was not justified by demand and cost, and the Society limited *Transactions* papers to relatively few selected *Proceedings* papers. However, it became too difficult to decide which papers merited the rating of "permanent reference value," and this economy survived for only three years until 1959.

Return to the publication of all *Proceedings* papers in *Transactions* during the period 1960–63 was equally unsatisfactory. The number of papers had risen to 330 per year, and the Society had to publish the 1961, 1962, and 1963 volumes of *Transactions* in five parts. The operation was incurring an annual loss of $75,000.

At that point, the Society decided to produce *Transactions* as a compilation of digests of all *Proceedings* papers, together with the complete "Annual Address of the President" and abstracts of the memoirs of deceased members. The paper digests, preferably prepared by the authors, were intended to be functional capsule presentations of the original paper, its discussion, and the closure. The abstract would meet the reference requirements of the searcher in most cases. Digests of all *Civil Engineering* articles were also included in the *Transactions* volume. One compact bookshelf reference then provided a key to an entire year of the Society's publications. The vast majority of *Transactions* purchasers favored this move.

By 1960, the technological "information explosion" had reached such proportions that the Society developed new measures to assist literature users. In 1963, staff developed a system of abstracts and keywords in accord with the *EJC Thesaurus of Engineering Terms.* This was followed, in 1966, by a new bimonthly *ASCE Publications Abstracts*—a compilation of information-retrieval cards for all *Civil Engineering* and journal papers, together with subject and author indexes. The reader could clip these for a personal reference file.

In 1970, ASCE, resorted to the electronic computer to best serve the information needs of the profession. A single input of information of all the abstracts and indexes that unify the many periodical publications of the Society was possible. Computerized abstracting, indexing, and typesetting were so efficient that the Society proposed in 1973 that a monthly *Civil Engineering Abstracts* be published jointly by all of the professional and technical organizations serving the various specialized areas of civil engineering practice. The idea was well received and appeared likely to develop successfully.

ASCE devotes a substantial segment of the operating budget to its publications.

Manuals of Practice

The Manual of Engineering Practice series was initiated in 1927 when the Board of Direction sought a publication format for a *Code of Practice*, practical ethical guidelines for the professional engineer. This was Number 1 of the long series of monographs specifically authorization by the Board of Direction.

The Technical Procedure Committee defined the Manual in 1930 as "an orderly presentation of facts on a particular subject, supplemented by an analysis of the limitations and applications of these facts." Also, it is "the work of a committee or group selected to assemble and express information on a specific topic." The Manual "is not in any sense a 'standard,' however; nor is it so elementary or so conclusive as to provide a 'rule of thumb' for non-engineers."

In 1962, the series was renamed Manuals and Reports on Engineering Practice. Manuscripts had to be published in a journal and reviewed by the entire membership in order to be published as a Manual. As of 1973, there were 53 numbered Manuals, of which seven were on professional subjects and the remainder on technical matters.

On several occasions, ASCE has collaborated with other technical and professional organizations to produce special reports. Typical of these are the following:

- Glossary: Water and Wastewater Control Engineering (with AWWA, American Water Works Association; WPCF, Water Pollution Control Federation; and APHA, American Public Health Association)
- Recommended Guide to Bidding Procedure on Engineering Construction (with AGC, Associated General Contractors of America)
- Water Treatment Plant Design (with AWWA and CSSE, Combat Service Support Element)
- Design and Construction of Sanitary and Storm Sewers (with WPCF)

ASCE has also published a wide assortment of special-purpose books of various kinds. Some of these, such as the newsletters of the Technical Divisions and the *Civil Engineering Research Letter*, provide administrative communication to technically oriented activities. Others relate to internal functions, such as the newsletters concerned with Sections and Student Chapters.

As early as 1871, a "List of Members" was furnished as a membership service. In 1882 the Constitution and Bylaws were included in this annual booklet. Lists of committees, award recipients, Local Sections, and student chapters were added as time passed, with more and more general information for members. In 1918, the publication was named the *Yearbook*. Beginning in 1950, the membership lists were published in a separate *Directory*, and the remainder of the material in a booklet titled the *Official Register*. For reasons of economy, the *Directory* became a biennial publication in 1956, and since 1962, members have paid for it.

The Board of Direction issues a report on the state of the Society at the Annual Meeting, as directed by the 1852 Constitution. From 1937 to 1949, the full report appeared in the *Yearbook*, although this was not required by the Constitution, and a "Secretary's Abstract" was carried in the *Official Register* from 1950 to 1955. The Board decreed in 1956 that an Annual Report go to each member, reasoning that an informed membership would be more interested, enthusiastic, and active. The new format was also an effective public relations tool, bringing widespread attention to the work of the Society on the part of nonmembers and other organizations, and served to promote membership. The cost was only about 25 cents per member.

The publications policy of the Society has overseen the printing and distribution of many documents of interest and value. The full range of miscellaneous publications extends from the Conference Proceedings volumes containing the papers from Specialty Conferences, to the ASCE Historical Publication Series, to the translation of the Russian monthly journal *Gidrotekhnicheskoe Stroitelstvo* (Hydrotechnical Construction).

Organization for Technical Activities

For almost 70 years, ASCE's efforts to advance the art and science of civil engineering were primarily the national meetings and Society publications. No attempt was made to administer technical activities under a formal organizational plan until well into the twentieth century. Early technical committees were mostly concerned with methods for testing engineering materials, and with the application of such test data to rational design.

The Committee on Tests of American Iron and Steel was created in 1872 when General William Sooy Smith proposed:

> That a committee of five be appointed to urge upon the United States Government the importance of a thorough and complete series of tests of American iron and steel, and the great value of formulae to be deduced from such experiments.

Although this committee was not successful in enlisting continuing support from Congress, it established an example for the many technical committees that were to follow.

The Board adopted a 1984 committee recommendation to join with Stevens Institute of Technology in founding a testing laboratory "for making complete and impartial tests of the characteristics, value and strength of materials used in the arts."

In 1875, it appeared that the aims of the committee had been realized when a Presidential Order was issued by the White House; in part, it said:

Executive Mansion

March 25, 1875

In pursuance of the 4th section of the act entitled "An act making appropriations for sundry civil expenses of the Government for the fiscal year ending June 30, 1876," ... a Board is hereby appointed, to consist of Lieutenant-Colonel T.T.S. Laidley, Ordnance Department, U.S. Army; Commander L.A. Beardsley, U.S. Navy; Lieutenant-Colonel Q.A. Gillmore, Engineer Department, U.S. Army; David Smith, Chief Engineer, U.S. Navy; W. Sooy Smith, Civil Engineer; Alexander L. Holley, Civil Engineer; R.H. Thurston, Civil Engineer, who will convene at the Watertown Arsenal, Massachusetts ... for the purpose of determining by actual tests, the strength and value of all kinds of iron, steel and other metals ... to prepare tables which will exhibit the strength and value ... for constructive and mechanic purposes ... and the building of a suitable machine for establishing such tests.

Mr. R. H. Thurston, Civil Engineer, is designated as Secretary of the Board at an annual compensation of twelve hundred dollars.

Actual traveling expenses, as provided by law, will be allowed the members of the Board.

U.S. Grant

However, funds for the testing operation were denied in the next session of Congress. The committee's determination is manifest in the following extracts from its report of April 1877:

The Committee ... [alludes] here, to the want of knowledge of the characteristics of the new varieties of iron and steel offered for our use ... to bring the mind of each member of the Society to a realization of the value of the knowledge which seems just within our grasp. Shall we fail to attain it?

There is not a member of Congress of the United States who cannot be reached through some member of our Society, who is personally acquainted with him. Let us make a vigorous effort at once to get Congress to repeal the legislation discontinuing the Board when the money already appropriated has been expended and to appropriate the money which the Board may need.

The testing machine erected in the Watertown Arsenal was turned over to the U.S. Ordnance Department in 1879. Although the 1881 appropriations bill contained a directive that the Chief of Ordnance "give attention to such programme of tests as may be submitted by the American Society of Civil Engineers," no systematic series of tests on engineering materials was ever accomplished. The analysis of iron and steel continued to receive specific committee attention in ASCE, however, until 1903.

Among other early outstanding technical committees were the

- Committee on Rapid Transit and Terminal Facilities, 1873;
- Committee on Railway Signals, 1874 and 1875;
- Committee on Cost and Work of Pumping Engines, 1874;
- Committee on Preservation of Timber, 1880;
- Committee on Proper Manipulation of Tests on Cement, 1898;
- Committee on Rail Sections, 1875;
- Committee on Concrete and Reinforced Concrete, 1905;
- Committee on Steel Columns and Struts, 1909;
- Committee on Materials for Road Construction, 1910;
- Committee on Bearing Values of Soils for Foundations, 1913;
- Committee on Stresses in Railroad Track, 1914;
- Committee on Bridge Design and Construction, 1921;
- Committee on Valuation of Public Utilities, 1910; and
- Committee on Highway Engineering, 1920.

Octave Chanute (President, 1891) served on at least four committees operating in 1875, and he was chairman of two of them.

Origin of the Technical Divisions

The 1919 report of the Committee on Development, which foresaw significant expansion in ASCE's professional activities, also envisioned growth in the technical area. It urged additional meetings and emphasis on the publication of papers at both national and local levels. The committee also proposed that standing advisory committees to the Board of Direction be appointed "to promote the study of important engineering subjects and ... outlining and coordinating the work of like committees of the Local Sections." The Board did not act on the report, but it was undoubtedly a factor in the broadened aims and responsibilities accepted by the Society in the 1920s and thereafter. On February 4, 1921, Director John C. Hoyt addressed the following proposal to the Board of Direction:

> With a view to increasing the value of the technical activities of the American Society of Civil Engineers ... [we propose] civil engineering be divided into several branches (for example, sanitary, hydraulic, structural, topographic, highway, railroad, irrigation, drainage, river and harbor improvement), and that a committee ... follow the work in each branch, such committee to be composed of men ... active in the branch. Following are possible functions of each such committee:
>
> 1. To keep in touch with the work that is being done throughout the country in its branch.

> 2. To advise the membership in a general way, through the *Proceedings*, of activities in its branch.
>
> 3. To solicit papers for presentation to the Society ... [so] that the Society publications may contain a record of important current activities in engineering, as well as a history of completed engineering work.
>
> 4. To keep in touch with the activities of other organizations [so] that duplication [is] avoided.
>
> 5. To initiate ... subcommittees ... within limited fields in the branch if there is demand for them.
>
> 6. To assist the Committee on Publication, especially by reviewing manuscripts submitted for presentation, or to suggest names for this function.
>
> 7. To study systematically the needs for subcommittees on research problems in the branch and to make appropriate recommendations to the Board of Direction.
>
> The above plan of procedure should tend to create an interest by the individual members ... and the committees could be composed of the younger men. ... It could also be extended to promote activities in the Local Sections, and might eventually lead to the formation of technical sections.

Two months later, the Board appointed a Committee to Promote the Technical Interests and Activities of the Society, under the chairmanship of Past-President Arthur N. Talbot. The Board adopted this body's report in January 1922, providing for a standing Committee on Technical Activities and Publications, with advisory subcommittees covering the principal branches of the civil engineering field.

No time was lost in effecting the necessary Bylaws amendments. When the Board approved the new committee and the new Technical Divisions on first reading, the officers were authorized to proceed without waiting for formal adoption at the next meeting. The Board acted in June 1922 to authorize formation of the Power Division, Sanitary Engineering Division, Irrigation Engineering Division, and Highway Division, in that order. The latter two authorizations were subject to the subsequent filing of the required petitions.

At this point, Acting Secretary Elbert M. Chandler cited 16 existing organizations with overlapping and duplicating interests in civil engineering, and he urged that action be taken to invite these bodies to become Technical Divisions of ASCE. A committee was authorized to promote such affiliations, but none were made.

As originally established, the Technical Divisions were administered by their own executive committees; they could enroll nonmembers of ASCE as affiliates, and they could assess dues to finance their operations. They were directly responsible to the Board of Direction. In 1926, the Board created a separate Committee on Technical Procedure to coordinate all technical affairs.

Table 6-1. *ASCE Technical Divisions, 1922–74*

	Year	*1973*	*1973 Committees*	
Technical Division	*Established*	*Enrollment*	*Number*	*Personnel*
Power	1922	2,992	13	73
Environmental Engineering	1922	10,031	20	168
Irrigation and Drainage	1922	4,660	22	129
Highway	1922	12,197	12	97
Urban Planning and Development	1923	9,215	22	129
Structural	1924	23,546	80	656
Waterways, Harbors, and Coastal Engineering	1924	4,159	30	156
Construction	1925	18,774	23	188
Surveying and Mapping	1926	3,731	17	97
Engineering Economics	1931	—[a]	—[a]	—[a]
Geotechnical Engineering	1936	15,247	24	179
Hydraulics	1938	11,029	39	220
Air Transport	1945	1,889	13	58
Engineering Mechanics	1950	6,098	18	150
Pipeline	1956	1,639	15	165
Urban Transportation	1971	1,027	13	56

[a]The Engineering Economics Division was dissolved in 1952.

Membership included the chairmen of the Publications Committee, Research Committee, and of all Technical Division executive committees, together with the President and Secretary of the Society and two members of the Board. Separate Technical Division dues were discontinued in 1943, when provision was made for the Divisions in the regular operating budget of the Society.

All the Technical Divisions formed between 1922 and 1974 are listed in Table 6-1. Their influence has been extended into the Local Sections by way of technical groups in the Sections, and each Technical Division has its own newsletter to further membership interest and communication.

The 1930 reorganization of the Society under the Functional Expansion Program classified all activities into three departments: Administrative, Professional, and Technical. The Technical Procedures Committee, Technical Divisions, and the Research Committee constituted the new Technical Activities Department. Following are extracts from the Expansion Program report:

> The technical functions ... are well performed through the ... Society committees and Technical Divisions. No change is proposed in them. ... The program provides ... for development ... along non-technical or professional lines.

This was the first official recognition of technical activities as apart from professional activities. In subsequent years, this dichotomy acquired a competitive aspect that extended to budgetary considerations, planning of meeting sessions, and even to the assignment of Board members to standing committees.

The Technical Department operated effectively until 1950, when another reorganization eliminated this arrangement altogether. However, Technical Division operations continued to be independent under a Committee on Division Activities that included Board of Direction members. A Committee on Technical Procedure included the chairmen of the Technical Division executive committees. A few years later, the Divisions were authorized, if they chose, to name a Board of Direction "contact member" to meet on invitation with the Division executive committee, but without vote. In 1963, a Director from the Board was designated as a full member of each Division executive committee. Such Board contact members had served on professional committees for many years.

In 1963 the Board created the Technical Activities Committee (TAC) to oversee all technical affairs of the Society. TAC was a counterpart of the Professional Activities Committee (PAC). The Technical Divisions were represented by their Board contact members in TAC, just as the various professional committees were represented in PAC. The Division executive committee chairmen were no longer directly involved in the general administration of technical activities.

For almost 30 years after they were authorized, the Technical Divisions operated separately, with only nominal collaboration in areas of overlapping interest. In 1959, there was an important change, when the Hydraulics, Sanitary Engineering, Irrigation and Drainage, Power, and Waterways and Harbors Divisions set up a Coordinating Committee on Water Resources to advance their mutual concerns. A similar Coordinating Committee on Transportation followed, bringing together the Air Transport, City Planning, Highway, Pipelines, and Waterways and Harbors Divisions; it was dissolved in 1962. Coordinating committees on Civil Defense (1960), Water Rights Laws (1964), and Structural Engineering (1967) also served effectively.

More formal liaison was introduced in 1965 with the formation of the Technical Council on Urban Transportation. Other Technical Councils soon followed: Ocean Engineering (1967), Aerospace (1970), Water Resources Planning and Management (1972), and Computer Practices (1973).

After 1968, the pendulum made another arc when the Professional Activities Study Committee brought about the extensive reorganization of professional activities (see Chapter 4). Even as this new mechanism was implemented, the Board of Direction in 1971 approved a restructuring of technical affairs on the recommendation of a Technical Activities Study Task Committee. This change grouped the Technical Divisions and Technical Councils into five "Management Groups," comprising Board contact members and nonmembers of the Board. The move added another level of administrative review. A proposal originating in a Technical Division or Council had to be cleared by the Management Group and then by TAC before it could be considered by the Board. At the same time, the change relieved the Board of routine operational oversight of the Technical Divisions and Councils, delegating this authority to TAC and the Management Groups.

"The technical and professional activities of ASCE are carried on in two separate spheres of the Society, united only at the Board of Direction level." This is the basic premise of a Task Committee on Cooperation Between PAC and TAC appointed in 1972 "to study cooperation and coordination between the Technical and Professional Activities of the Society." The 1973 report of the task force presented recommendations for improving communication and interaction between the two areas in the Board of Direction and staff, in meetings, in committee work, and in continuing education. The report stated that "It is not necessary that TAC and PAC have the same organizational setup ... [but] it is essential that TAC and PAC activities be thoroughly coordinated and tied more closely with local section programs."

Time has proved Director Hoyt's suggestion to the Board of Direction in 1921 to have been a major contribution to civil engineering, with an impact on the entire engineering profession. His concept of the Technical Divisions as a means of advancing the art was fulfilled to the letter. In a broader sense, the movement had two opposing effects on the turn-of-the-century trend toward proliferation into specialized technical organizations. On the one hand, the Technical Divisions slowed the trend by supplying the need for identification and services in some specialties; on the other, several technical associations were actually created by ASCE members as a by-product of their activities in the Divisions.

Development of Civil Engineering Standards

The coordinated programs in ASCE directed toward engineering standards and research evolved from the same special committees that preceded the Technical Divisions. With a few exceptions, the standards activities have been completely cooperative with organizations whose main purpose is establishment of engineering standards. ASCE has always maintained that standards should be founded on the broadest possible base of acceptance, rather than a single organization.

The independent activities of ASCE in standards development extended over the early years of the Society, when there were few other engineering societies. In the period 1875–95, important committee reports relating to standardization were produced on such subjects as tests of American iron and steel, railway signals, standard rail sections, uniform systems of tests on cement, compressive strength of cements and cement mortars, uniform methods of testing materials, and preservation of timber. Two early ventures of ASCE into the realm of standards merit documentation were the Society's role in connection with the metric system of weights and measures and its role in establishing the present system of standard time.

Consideration of the Metric System

The concern of ASCE with regard to adoption of the metric system of weights and measures goes back to 1876, when Clemens Herschel (President, 1916) offered this resolution:

> *Resolved:* That the American Society of Civil Engineers will further ... the adoption of the metric standards in the office of weights and measures at Washington, as the sole authorized standard of weights and measures in the United States; that the Chair appoint a committee of five to report to the Society a form of memorial to Congress in furtherance of the object expressed.

The Board of Direction submitted the resolution to letter ballot of the membership, and it was approved by a vote of 138 to 73. The Society appointed a Committee on Metric System of Weights and Measures and decided that authors of papers presented before the Society should include the metric equivalents of all weights and measures in those papers.

The Congressional "memorial" drafted by the committee failed to be adopted by a letter ballot in 1877, and the committee was not continued. An 1878 resolution adopted by membership referendum postponed further consideration of the metric system. By 1881, only two authors had complied with the resolution calling for the inclusion of metric equivalents in ASCE papers, and the policy was rescinded.

In 1907, the Board appointed a special Committee on Status of the Metric System in the United States. The movement was still strongly opposed, however, and the Board discharged the committee in 1910, without approving its recommendation that the report be sent to all members of Congress and the New York State legislature.

The Publications Committee decided in 1917 not to reconsider an ASCE requirement that metric equivalents be used in its technical publications.

Eighty-one years after ASCE resolved that "further consideration of the metric system of weights and measures be postponed," the Board of Direction acted in October 1959 to reaffirm the 1876 policy that "the Society further, by all legitimate means, the adoption of the Metric Standards ... as the sole authorized standard of weights and measures in the United States." In 1969, the Committee on Standards created a Task Committee on Metrication to study the effects of conversion on civil engineering and the construction industry in the United States.

The task force recommendations resulted in the 1971 adoption by the Board of Direction of the following:

1. To support conversion to the International System (known as SI).
2. To recommend that no additional metric units be incorporated into the SI System.
3. To begin immediately the use of dual units in all ASCE publications.

4. To adopt the ASTM Metric Practice Guide.
5. To recommend the use of SI units in all cartographic and geodetic projects.
6. To recommend revision of pertinent publications using SI units.
7. To recommend that instruction in SI units be initiated immediately in the teaching of civil engineering subjects.

In 1972, the Society published a brochure, "Metrication and the Civil Engineer," for use as part of an information packet.

ASCE Fosters Uniform Standard Time

The most important and exciting venture by ASCE into the realm of standards was the Society's effort toward a uniform system of standard time to replace the existing unrelated local time systems throughout the world. It began when Sandford Fleming, a Canadian member, presented a paper "On Uniform Standard Time for Railways, Telegraphs and Civil Purposes Generally" at the thirteenth Annual Convention in June 1881.

Fleming described the time notation situation then prevalent, which was the cause of great confusion, especially in North America:

> According to the system of notation we have inherited from past centuries, every spot of earth between the Atlantic and Pacific is entitled to have its own local time ... each locality ... may insist upon its railway and its other affairs being governed by the time derived from its own meridian. The smaller and less important localities ... have found it convenient to adopt the time of the nearest city. The railways have laid down special standards which vary, as has been held expedient by each separate management. In the whole country [both Canada and the United States] there is ... an irregular acknowledgment of more than one hundred of these artificial and arbitrary standards of time. The consequences ... are felt by every traveler, and in an age and in a country when all, more or less, travel, the aggregate inconvenience ... is very great and it will be enormously multiplied as time rolls on. If the system already results in difficulties ... which often occupy our courts of law, which ... too often ... cause loss of life, what will be the consequences in a few years, when population will be immensely increased and travel and traffic ... multiplied, if no effort be made to effect a change?

The problem had already been recognized by the American Meteorological Society, the Imperial Academy of Science of Russia, the Royal Society of London, and the Canadian Institute. The U.S. and Canadian organizations jointly recommended a system of 24 standard meridians, with the prime or zero time meridian to pass near the Bering Strait, 180 degrees from the Greenwich Observatory. They also recommended consecutive numbering of 24 hours, instead of the 12 hours ante meridiem and post meridiem (a.m. and p.m.)

Fleming sought an expression of support from ASCE for these proposals, saying that such an endorsement "must carry with it great weight, and will exact respect in every quarter." His eloquent plea found receptive ears, and he was appointed chair of a new ASCE study group.

At its next meeting in June 1882, the Society affirmed the need for a standard time system, and it endorsed the idea of the 24-hour universal day to be reckoned from a prime meridian of longitude. At the January 1882 Annual Meeting, the committee requested authority "to invite the cooperation of other scientific associations ... in the furtherance of this important object, and that all such societies and government departments interested be invited in the name of the Society to attend a general convention to meet at New York or Washington ... for the purpose of determining the Time System advisable to adopt." These recommendations approved, and the President of the Society was authorized to invite representatives from Canada and Mexico, as well as appropriate state and federal agencies, to the proposed conference.

Chairman Fleming and his associates lost no time in developing a brochure outlining the plan, and the Society circulated this material among railroad and telegraph company officers, leading scientists, and public officials. The response was almost unanimously favorable. The Society authorized the committee to petition Congress to take the necessary steps to establish a prime meridian.

Such momentum developed that there was no patience for the ponderous movement of Congress. In his report at the January 1884 Annual Meeting, Chairman Fleming stated:

> On the 11th October last, the railway authorities met in convention at Chicago and ... decided to adopt the hour standards, and they fixed upon the 18th of November (1883) as the day when they would generally begin to operate their lines by the hour meridians. The public with great unanimity acquiesced to the changes. It is now generally and universally admitted to be a great public boon.

Thus did the present system of Standard Time come to North America, with its Atlantic, Eastern, Central, Mountain, and Pacific time zones reckoned from the Greenwich meridian. This action did not, however, encompass the so-called 24-hour notation.

Meanwhile, Congress had responded to the ASCE petition by adopting on August 3, 1882, a joint Resolution authorizing the President "to call an International Conference to fix on and recommend for universal adoption a common prime meridian to be used in the reckoning of longitude and the regulation of time throughout the world."

The International Meridian Conference was held in Washington in October 1884, with representatives of 26 nations participating. The conference resulted in essential acceptance of the uniform standard time advocated by ASCE, as manifest in "the system of regulating time which has been adopted with signal success in North America." The following resolutions were adopted:

1. That it is the opinion of this Conference that it is desirable to adopt a single prime meridian for all nations, in place of the multiplicity of initial meridians which now exist.
2. That the Conference proposes to the Governments here represented the adoption of the meridian passing through the centre of the transit instrument at the Observatory of Greenwich, as the initial meridian for longitude.
3. That from this meridian longitude shall be counted in two directions up to 180 degrees, east longitude being plus and west longitude minus.
4. That the Conference proposes the adoption of a universal day for all purposes for which it may be found convenient and which shall not interfere with the use of local or other standard time when desirable.
5. That this universal day is to be a mean solar day; is to begin for all the world at the moment of mean midnight of the initial meridian, coinciding with the beginning of the civil day and date of that meridian; and is to be counted from zero up to twenty-four hours.
6. That the Conference expresses the hope that, as soon as may be practicable, the astronomical and nautical days will be arranged everywhere to begin at mean midnight.
7. That the Conference expresses the hope that the technical studies designed to regulate and extend the application of the decimal system to the division of angular space and of time shall be resumed, so as to permit the extension of this application to all cases in which it presents real advantage.

Two months after the Washington conference, on January 1, 1885, the hour zone system was adopted at the Greenwich Observatory. This began a slow but steady movement toward worldwide acceptance. In 1891, Chairman Fleming reported:

> The hour zone system has been adopted for ordinary use in portions of the three continents of Asia, Europe and America. In 1887 an imperial ordinance was promulgated directing that on and after the first day of January of the year following, time throughout the Japanese Empire would be reckoned by the third hour meridian. The reckoning in England and Scotland is by the twelfth hour meridian; in Sweden the eleventh hour meridian is the standard, and quite recently it has been resolved in Austria–Hungary to be governed by the same meridian. Efforts are now being made to follow the same course in Germany and in other European countries.

Another recommendation of the international conference—the "24-hour clock"—was not to meet with such success. This reform would have been

Figure 6-2. *The 24-Hour Clock*

Source: ASCE *Proceedings*, June 1884, p. 77.

effected by universal use of the watch dial illustrated in Figure 6-2. The concept was applied to the publications and meetings of ASCE in 1883. Also known as the "Universal Day," the notation was subsequently adopted by the Greenwich Observatory, the railways of Canada, the Eastern Telegraph Company (serving parts of Asia, Africa, Australia, Europe, and New Zealand), and by several national governments. Because one of the international conference recommendations referred to the possibilities of a decimal time system, the issue became controversial. While apparently favored by a wide majority of railroad managements in the United States, they did not adopt the 24-hour clock as they had the hour zone system in 1883.

When President Grover Cleveland in his 1886 message to Congress urged adoption of the recommendations of the International Meridian Conference, there were high hopes that Congress would give legal sanction to these proposals. Chairman Fleming said it would be appropriate and feasible for the Universal Day of 24 consecutively numbered hours to be formalized on January 1, 1900. But this was not to be.

Several surveys and mailings of brochures aroused sufficient interest on the part of railroads, telegraph companies, and the public that in 1890, the Board of Direction produced a "memorial" urging Congress to adopt the recommendations of the 1884 international conference. The resolution, approved by a mail ballot of the membership, was conveyed to President Benjamin Harrison and to both Houses of Congress. In January 1891, bills were introduced in both Houses "respecting the reckoning of time throughout the United States," setting forth the principles of time reform advocated by the Society. Neither bill was passed.

Procrastination by Congress did not deter progress elsewhere in the world. Belgium, the Netherlands, and Germany adopted the hour zone system; India, the 24-hour notation. France would not accept the Greenwich Prime Meridian, but it adopted the reckoning of Paris as the time for the entire nation—a difference of only nine minutes from Uniform Standard Time.

The issue came back to Congress in 1896, in connection with the International Conference recommendation that astronomical time be abolished and replaced by civil time in nautical almanacs for the purpose of navigation. Six nations publishing nautical almanacs were prepared to accept this provision, but only upon condition that the United States would do so simultaneously. A most convincing resolution by the Society again petitioning Congressional approval of the International Conference findings was no more successful in overcoming the intransigence of Congress than its predecessors.

After these disappointments, the efforts of Chairman Fleming and his committee centered on the 24-hour notation system, but with a gradual decline in emphasis. In 1899, a small but vocal band of members protested the continued use of the 24-hour system in the notices of the Society because it was "not likely to accomplish the end for which it was adopted." The outcome of this discussion was an action to request the Committee on Uniform Standard Time to make its final report at the next Convention.

Chairman Fleming made his last report in a letter dated January 13, 1900. He reviewed the accomplishments of his committee since its appointment in 1881, noting that the hour zone system of Standard Time was by now effective on five continents. Although disappointed in the limited acceptance of the 24-hour clock, he still expressed hope:

> The American Society of Civil Engineers took a leading part in initiating Standard Time, and it has continuously stimulated the development of a great reform in time-reckoning, not on this continent alone, but throughout the world. ... At one time some were sanguine enough to think it possible ... on the opening of the coming new century, but whether then or later, I am satisfied that like the Gregorian reform, the modern time-reform must in the end become an accomplished fact.

With this report, 19 years of dedicated, significant activity was terminated by a terse expression of thanks to the committee. Sandford Fleming was chairman from 1881 to 1900. Others who served for 19 years were Thomas Eggleston, Charles Paine, and John M. Toucey. Theodore G. Ellis, Theodore N. Ely, J. E. Hilgard, and Frederick Brooks served for shorter periods. A more fitting acknowledgement of their contribution is the Society's action in June 1883:

> *Resolved*, That the American Society of Civil Engineers hereby acknowledges the extent and value of the work accomplished to date by the Committee on Uniform Standard Time, and tender to that Committee hearty thanks and earnest congratulations for the diligence and the intelligent and fruitful labors of which the results have been so well exhibited.

Evolution of "American Standards"

By 1916, engineering standards activities had become so complicated that ASCE, AIME, ASME, AIEE and ASTM (ASTM had been founded in 1898) collaborated to form a joint Committee on Organization of an American Engineers Standards Committee. The state of affairs at the time is summarized in the preamble of that committee:

> At the present time many bodies are engaged in the formulation of standards. There is no uniformity in the rules ... in the different organizations; in some cases the committees ... are not fully representative, and ... they do not consult all the allied interests. The present custom results in a considerable duplication of work, and ... in some fields several "standards" proposed for the same things differ from each other only slightly. ... It is very much more difficult to obtain agreement between the proposers of overlapping standards after they have been published than to get the proposers to agree before they had committed themselves publicly.

The joint study resulted in a recommendation for the creation of a permanent American Engineering Standards Committee, with representatives from the five sponsoring societies and others admitted later. Specialist "Sectional Committees" would develop specific standards, with input from interested "Cooperating Societies," including government agencies and others. Standards approved by the main committee were to be designated first as "Recommended Practice," and, when suitably established, as an "American Standard."

The American Engineering Standards Committee (AESC) was established in October 1918. Its membership was soon expanded to include government and industrial bodies concerned with standards, and it performed an important service in coordinating the work of the standards—producing committees of the many "sponsor" bodies.

In 1928, AESC was reorganized "to permit broader participation by trade associations and others having an important concern with National standardization work," and it became the American Standards Association (ASA). Procedures in ASA were broadened and the Sectional Committees modified to include representatives of all interested entities. By 1930, ASA had 320 projects on its agenda, and 155 American Standards were in use. The association was affiliated with the International Organization for Standardization (ISO). ASCE participated through its representatives in 22 Sectional Committees with a broad range of subjects.

ASCE Standards Activities in Recent Years

During the next 35 years, ASCE standards activity was conducted through its cooperation with ASA and ASTM. From time to time, the Society collaborated with bodies such as the U.S. Bureau of Standards and various industrial associations on special problems.

In the early 1950s, some of the ASA Sectional Committees were not as active as they might have been, and ASCE revised the means for selecting and reviewing appointments to these committees. The Society appointed a Committee on Standards in 1956, to work closely with the Technical Divisions and serve the Board of Direction in an advisory capacity. Several Divisions set up their own standards committees to support the Society committee, to involve effective personnel, and to expedite the operations of ASA with civil engineering standards. In 1961, the Society accepted an invitation to name representatives to the U.S. National Committee of the ISO.

ASA underwent a second major reorganization in 1966, becoming the U.S.A. Standards Institute (USASI). ASCE continued to be represented through the Member Body Council, and carried on its participation in some 35 Sectional Committees and Correlating Groups. In 1970, USASI was renamed as the American National Standards Institute (ANSI).

The standards program assumed new importance in the mid-1960s with a rising interest in building codes. The U.S. Chamber of Commerce was represented, as were other interested professional organizations, in the Model Code Standardization Council. In 1968, ASCE's Committee on Standards joined the National Fire Protection Association and the American Institute of Architects in exploring the best approach to developing a model national building code. Although this issue has not been resolved, these studies instigated several actions at the national level toward consistency in building codes.

In 1974, about 150 members represented the Society in some 75 standards and codes activities, including seven Technical Advisory Boards and 28 committees of ANSI, 17 committees of ASTM, three U.S. National Committees of the ISO, various committees of the U.S. Bureau of Standards and other government and private organizations, and in the Model Codes Standardization Council. ASCE was cooperating with other major engineering societies in support of federal legislation on metric conversion. The recently enacted regulations of the federal Occupational Safety and Health Act conflict with long-established local building codes, and these divergences were being reviewed by the Task Committee on Building Codes. New task committees on nuclear standards and on building code requirements for excavations and foundations were also operating under the aegis of the Committee on Standards.

Advancement of the Art through Research

Research services in ASCE have generally been carried on under the direction of the Technical Divisions, with emphasis on efforts in the most urgent problem areas. The Society has sought to provide two-way communication between civil engineering practitioners and the research institution, both in identifying research needs and in making the results of research available. It has a catalyst for effective interaction between science and art.

ASCE's Technical Divisions and standards activities include a number of very early ad hoc technical committees. Some of these were engaged in testing engineering materials; others addressed reviews of the state of the art in various aspects of civil engineering practice. While these efforts do not represent fundamental research, they are examples of the kind of applied research that was needed in the latter part of the nineteenth century.

Arthur N. Talbot (President, 1918), in a review of early research activities (*Transactions*, Vol. 86, p. 1280), cited as particularly significant the committee work on uniform tests of cement (1912), concrete and reinforced concrete (1917), steel columns and struts (1919), bearing values of soils, and stresses in railroad track. Past-President Talbot provided much of the leadership toward the formation of research programs in the Society. He considered that a national engineering society should be a "stimulator of research, and of the progress of engineering science."

ASCE and the other Founder Societies consolidated research-related efforts in the 1914–20 period when Ambrose Swasey, a Past-President of ASME, gave endowments of $500,000 to the United Engineering Society (later United Engineering Trustees) for "the furtherance of research in science and engineering, or for the advancement in any other manner of the profession of engineering and the good of mankind." This was the beginning of the Engineering Foundation, which has achieved a distinguished record in the advancement of engineering research in all fields. Administered by representatives of the Founder Societies, the foundation has financed—wholly or in part—many millions of dollars of important technological pioneering. ASCE has been a strong supporter and a frequent applicant for research funding. ASCE projects in the early years of the foundation included research on concrete and reinforced concrete arches, steel columns, soils and foundations, and arch dams.

When in 1916 the National Academy of Sciences formed the National Research Council (NRC) as a wartime advisory body of scientists and engineers, the Engineering Foundation offered the funding and staff needed until NRC was permanently authorized in 1918. All of the Founder Societies are represented in the NRC Engineering Division.

In 1920, ASCE identified research activities as an area of concern. In that year, the Engineering Foundation invited ASCE to name representatives of its Research Committee "to cooperate with other Founder Societies and with Engineering Foundation as to the best possible … research." The Board of Direction designated the chairmen of the Committee on Bearing Power of Soils, the Committee on Stresses in Railroad Track, and the Committee on Specifications for Bridge Design and Construction to serve as a Conference Committee on Research.

Just two years later, the committee was expanded to nine members, and given advisory status by the Board of Direction regarding the appointment of new special research committee. In 1923, the Society gave the Research Committee direct supervision over nine special committees—those on Bearing Values of Soils, Stresses in Railroad Track, Concrete and Reinforced Concrete Arches, Flood Protection Data, Hydraulics Phenomena, Impact in Highway Bridges, Irrigation Hydraulics, Steel Column Research, and Stresses in Structural Steel.

This arrangement prevailed for the next 15 years, in which period the Technical Divisions came into being and gained in strength. In 1938, the Society abolished the Committee on Research, and assigned its committees to the Technical Divisions. When it became apparent that coordinating Division research operations was essential, the Committee on Research was reconstituted for that purpose in 1946.

In 1947, a committee in the Structural Division proposed the Reinforced Concrete Research Council (RCRC), an independent entity supported by Engineering Foundation with ASCE sponsorship. With the cooperation of the American Concrete Institute, this council introduced the ultimate-strength design concept, which has been widely applied in structural engineering.

In 1958, the Research Committee held the first of a series of conferences highlighting research opportunities. In 1959, the committee initiated ASCE administration of research councils such as RCRC, formed for research in a particular field by all those interested in the problem, with funds contributed by interested groups. A new Research Council on Pipeline Crossings of Railroads and Highways was soon formed, and the RCRC transferred from the Engineering Foundation to the administrative umbrella of ASCE. These and other research councils created during the next 15 years became a major segment of ASCE research activity.

The year 1962 brought important developments. With the aid of a $10,000 grant from the Engineering Foundation, ASCE added a professional Research Manager to the staff. At the same time, the Research Committee broadened its program to encompass three new objectives:

- to clarify the Society's research goals;
- to upgrade and improve the image of civil engineering research; and
- to obtain increased institutional support for civil engineering research.

Under these guidelines, the new ASCE Research Department began with an unprecedented level of research activities in the Technical Divisions. Research conferences were part of the new effort, and a 1963 publication, "Advancing Civil Engineering Techniques Through Research," was another catalyst. In 1962, the Society established the ASCE Research Fund for special research purposes. In 1965, a Task Committee on the Study of National Requirements for Research in Civil Engineering requested input from every Technical Division. Each year brought at least one new research council.

The dismantling of the New York World's Fair in 1966 afforded a unique opportunity for destructive testing of structures. With funding of about $250,000 from several foundations and government agencies, ASCE joined with NRC's Building Research Advisory Board in this effort, which led to explorations of other such testing opportunities.

The Society's policy has been to aid to financing civil engineering research in qualified institutions when possible and to accept grants for direct research project administration only when preferred by the grantee or otherwise justified. A typical example was the 1966 Combined Sewer Separation Project, funded by the federal Water Pollution Control Administration. In 1969, ASCE gave the Research Committee broad authorization to engage in projects in which the Society would accept research funds for direct administration.

The public concern with the social impact of technology was reflected in ASCE research operations in the early 1970s. One example was the 1971 Conference on Goals of Civil Engineering Research—Its Responsiveness to the Needs, Desires and Aspirations of Mankind. The Society also undertook the project "Case Studies of the Impact of Civil Engineering Projects on People and Nature." Other Technical Division and Engineering Foundation conferences also pursued this theme.

The Research Committee initiated several measures to stimulate general membership interest in research. These included five annual research prizes, which were initiated with a modest stipend in 1946 and were endowed in the name of Past-President Walter L. Huber in 1965. The Society also approved an annual research luncheon to draw attention to the winners of these prizes. From 1961 to 1966, ASCE funded a $5,000 Research Fellowship each year. Other research awards are the O.H. Ammann Research Fellowship in Structural Engineering and the Raymond C. Reese Research Prize in Structural Engineering.

The Society's 1971 report, "Research Needs in ASCE Relevant to the Goals of Society," estimated such needs to require funding on the order of $3 billion annually for the next 10 years. Research councils administered by ASCE include Pipelines, Reinforced Concrete, Air Resources Engineering, Urban Hydrology, Environmental Engineering, Coastal Engineering, Urban Transportation, Performance of Structures, Urban Water Resources, Construction, Expansive Soils, Underground Construction, Structural Plastics, and Computer Practices.

ASCE also cooperates with many other organizations in their research councils and related activities. The Column Research Council, the Riveted and Bolted Joints Research Council, and the Steel Structures Painting Research Council are just a few examples. In the decade ending in 1973, collaboration with the Engineering Foundation was particularly productive. More than 50 of the Foundation's conferences were on civil engineering research topics, and projects of the ASCE Technical Divisions and research councils were being widely supported.

In 1974, more than 400 members of the Society were involved in 40 research committees and 13 research councils. Their projects—principally initiating and stimulating study in new areas of research—involved more than $500,000 of outside funding.

Sound research programs are an essential element in the pursuit of a learned art in the modem world. While not all the early leaders of ASCE agreed on the Society's role relating to the "professional improvement of its members," there was unanimity of opinion that advancement of technology was fundamental. The Society's development of meetings, publications, technical committees, standards, and research services came about more easily than did its growth in the professional domain. All the benefits furthered by ASCE cannot be enumerated here. Let it be recorded that the American standard of living—unmatched in the world—has been made possible at least in part through the efforts of the American Society of Civil Engineers.

CHAPTER 7

In the Spirit of Public Service

From the beginning, the American Society of Civil Engineers had a strong sense of responsibility for the safety, welfare, rights, and financial interests of the public. This is evident in the actions, writings, and biographies of early leaders such as Laurie, Craven, Kirkwood, McAlpine, Chanute, Adams, Jervis, Allen, Fink, Welch, Chesbrough, and many others.

The Constitution of ASCE does not state or even imply that the Society was to advocate the protection or advancement of the public welfare. Within its first 10 years, however, ASCE made several ventures into the domain of public affairs relevant to the "advancement of the profession." Conservative members questioned these actions, and policy developed slowly.

After its first two decades, ASCE involvement in public affairs increased. The Society gave advice on major public works programs and conservation issues to Congress—sometimes by invitation and sometimes not. Similarly, the Society recommended administrative procedures to government bureaus employing civil engineers. During both world wars, ASCE offered assistance to federal authorities on technical manpower problems, wartime construction, and postwar planning. Through the years, the Society has offered advice and guidance to government officials and agencies on natural resources and environmental quality.

Public Affairs and Public Service

It was not until 1919 that public service was identified as an institutional responsibility. When Chairman Onward Bates presented his 1919 Committee on Development report, the Board of Direction endorsed the premise "that the time has now come when this society should adopt the principle of becoming an active national force in economic and industrial and civic affairs." The Board authorized an ASCE Committee on Public Relations in 1921, which was intended to advise the Board on "matters of public policy and professional relations." In 1930, the Committee's functions were divided between a Committee on Legislation and a Public Educational Committee, both parts of a new Professional Department. The Committee on Legislation was charged "with being alert in the matter of legislation, other than registration, and to advise the Board promptly on needful decisions or actions." After three years, the Board discharged the committee and assumed their duties.

Identifying, evaluating, and obtaining informed consensus on legislative issues is properly a staff responsibility. These duties proved too much for the Board, and the cycle was repeated when, in 1946, the Board appointed a Committee on National Affairs to "advise and guide the Society in activities regarding all national legislation concerned directly with the welfare of the profession." Again, this was too much to expect from volunteers, and the committee was discharged in 1954.

In 1940, before the entry of the United States into World War II, the Committee on Society Objectives recommended that a member of the staff "spend as much time as necessary in Washington, particularly during the next few months, to look out ... for the interest of civil engineers. "High interest in the developing National Defense Construction Program resulted in the April 1941 designation of a full-time staff representative in Washington, only for the duration of the national emergency.

When the American Engineering Council closed its Washington office in 1941, ASCE invited the other Founder Societies to take part in a joint venture because of the national emergency and the need for a "measure of solidarity and concurrent action through representation at Washington of the four Founder Societies." AIME, ASME, and AIEE all declined the invitation, and ASCE's Washington Field Office continued for 14 years. Four staff members successively filled the Washington post, all civil engineers experienced in the legislative process, and at least two were registered lobbyists. They focused on legislation related to civil engineers in military service during World War II, to public works planning and construction programs, and to national policy on military training and on peacetime reconversion.

The representative also worked with federal agencies on regulations and procedures regarding engineering and construction, contracts, the civil service, research activities, and education. The office maintained liaisons with the American Institute of Architects (AIA), Associated General Contractors of America (AGC), and other engineering bodies based in Washington, and it cooperated with the Engineers Joint Council (EJC).

By 1955, the economics of the Washington office were no longer favorable, and the office was closed in 1956. From 1955 to 1972, staff handled all public affairs activity with assistance from several sources. ASCE engaged a

Washington law firm to screen all national legislative proposals and to alert Society headquarters to bills that might be of interest. About 150 to 200 bills were referred annually, and the Society took action on 25 to 40. Legislative concerns were also brought to attention by members, Local Sections, technical or professional committees, or other societies. The editors of *Civil Engineering* retained a professional journalist to write a monthly Washington news column for the magazine.

When a situation or problem was judged appropriate for ASCE action, officers tried to determine the Society's official position on the matter. If there was none, they sought direction from the Board of Direction, the Technical Divisions, or a special committee. The Water Policy Committee, Land-Use and Environmental Systems Policy Committee, and Transportation Policy Committee were particularly helpful.

Society policy, until 1922, implicitly represented membership consensus. From that year on, this was affirmed by a membership referendum. The process was questioned from time to time, as in 1890 when the Society wanted to adopt the 24-hour clock as the standard for time notation. Extensive debate ended with a call for a letter ballot, which supported the proposal. Constitutional changes in 1922 gave the Board of Direction broader powers. The changes made for more timely action, occasionally at the expense of clear-cut majority agreement on controversial questions.

A 1939 policy resolution dictated that no pronouncement on national policies was to be made by a committee or Technical Division without prior clearance by the Board of Direction.

Once the Society determined its position on a legislative proposal, the staff offered comments to sponsors or to relevant legislative committees. Frequently, the staff arranged for competent witnesses to appear before such committees. When legislation was enacted, the ASCE office offered advice to the appropriate officials and agencies concerning appointments, regulations, or procedures that might have engineering implications.

A provision in the Internal Revenue Service (IRS) regulations exempts educational and scientific organizations from federal income taxes only if no "substantial part" of the activities of such organizations "be devoted to the carrying on of propaganda or otherwise attempting to influence legislation." In deference to the IRS rule, ASCE has limited legislative activity to measures with a clear-cut public interest dimension, acting as competent adviser rather than as lobbying agency. Some of the bills acted upon were more or less self-serving, of course, but action was deemed justified on the premise that every nation needs a strong engineering profession to develop its natural and human resources properly. This logic might be paraphrased as "what is good for engineering is good for the country!"

Members were not exhorted to "write their Congressmen," although they were free to endorse the position of the Society as individuals. ASCE avoided any political-action techniques and any kind of activity in political campaigns.

The National Capital Section of ASCE plays an important role in the legislative public interest area. The Section had always maintained close liaison with the headquarters office, and it frequently cooperated in obtaining information and in making personal contacts on legislative and bureaucratic matters. Public affairs at the state and local levels have been almost entirely carried on through the Sections of the Society and their Branches.

Society policy has also cooperated with other professional and technical engineering organizations in public affairs of mutual concern.

The sweeping reorganization of ASCE professional activities in 1972 resurrected two operations related to public service, a general committee and the Washington Field Office. Actually, not one but two committees were created, both within the new Member Activities Division.

The Committee on Legislative Involvement was charged

> to stimulate and encourage participation by individual members in legislative activities on local, state and national levels. It shall inform itself on legislative problems and assist in developing Society policies. ... It shall ... monitor proposed legislation ... to determine its possible effect on Society activities and recommend corrective actions where necessary."

The Committee on Public Affairs was given an equally wide ranging assignment:

> The Committee shall encourage professional involvement by members of the Society as concerned citizens in local, national and international affairs. This will include assistance to government agencies and community action groups. It shall find ways for the civil engineering profession to better serve the public in the improvement of man's welfare and his environment.

These purposes went far beyond those of the two predecessor committees of the 1930s and 1940s. The new goals were ambitious but commendable, and a capable professional staff also promised success.

The Washington office reopened in 1972, this time staffed by an attorney knowledgeable in the ways of the national capital. Close rapport was promptly established with AIA, the National Society of Professional Engineers, the American Consulting Engineers Council, AGC, and other Washington-based organizations. The operation's early successes included professional contract negotiation legislation and with pensions.

Public affairs activity in the Sections increased, with a "Legislative Involvement Handbook," released in 1972, furnishing guidance in "organization and procedures to assist members involved in legislative activities at the Local Section level." Early in 1973, an ASCE conference encouraged Sections to assume leadership on state laws requiring professional negotiation for engineering services for public works instead of competitive bidding.

Altruism or Self-Interest?

Did the trend within ASCE to emphasize public service stem from self-interest or altruism? ASCE always sought greater professional status, public respect, and a higher order of economic welfare for the civil engineer. As time passed, a growing sense of public responsibility became evident.

The 1875 report of the Committee on Policy of the Society observed that:

> The influence of the Society upon the public is unfortunately not very rapidly developed by the mere professional character of the members, although this is the ultimate basis of its influence and usefulness. There must be some connecting links between the Society and the public.

At that time, such "links" consisted of publications and special committee studies.

In 1919, the Committee on Development stated: "The engineering profession owes a duty to the public which ... can best be discharged by every engineer in the civic work of his community."

The very purpose of the Functional Expansion Plan in 1930 was "to promote the desirable functions of the Society to best serve the public and the membership." The 1938 Committee on Professional Objectives also referred to "the responsibility of the engineer to society."

The 1962 Committee on Society Objectives and Tax Status considered modifying the objective of ASCE in Article I of the Constitution to stress "service in the public interest." The amendment was not recommended, however. The committee concluded that the principle of "by their works you shall know them" was of more consequence than a change of wording.

As a result of this committee's input, the preamble to ASCE's Code of Ethics changed in 1962 to recognize the public interest. This factor became thereafter a basic criterion in the evaluation of all Society programs.

In 1973, ASCE decided the first goal of the Society would be: "To provide a corps of civil engineers whose foremost dedication is that of unselfish service to the public." The self-serving enhancement of professional standing and economic welfare—although given emphasis in the Society's agenda—was to share priority in ASCE with service on behalf of the public.

There was never any membership groundswell of concern for the public welfare. When Presidential Nominee Mason G. Lockwood, sought guidance in 1955 from Local Sections about Society programs, there was strong demand for attention to the status of younger members, economic betterment, public relations, and membership apathy, but there was no mention whatever of services in the public domain.

There have always been countless members of ASCE who were outstanding as public-spirited citizens, more than generous in their contribution of time and talent to the public weal. Former President Herbert Hoover heads the long list.

There have been hundreds of public service contributions, some by formal action of the Board of Direction and others on the initiative of the staff. Some noteworthy examples follow, in public safety, public policy, wartime services, service to government agencies, and public communication.

Public Affairs Achievements

In its early years, ASCE analyzed national disasters relating to civil engineering. The Society appointed a committee, in 1892, to urge upon the members their duty to present papers recording "for the benefit of all not only their successful works but the experience not infrequently acquired by failure." As early as 1869, there was concern about the safety of bridges, which were failing at a rate of 25 or more annually. At the 1873 Annual Convention, the Board resolved:

> In view of the late calamitous disaster of the falling of the bridge at Dixon, Ill., and other casualties ... a committee ... be appointed to report at the next Annual Convention the most practical means of averting such accidents.

The committee, chaired by James B. Eads, thought it was "our duty as a Society to establish in a few general terms—such as can be readily embodied in a law—a standard of maximum stresses and a table of least loads for which bridges should be designed."

The committee's final report was published in 1875. There was disagreement, both on technical details and on the propriety of the Society's preparing a model bridge law. Although never formalized in law, the findings provided badly needed guidance for bridge designers.

During discussion of the Ashtabula, Ohio, bridge tragedy in 1877, Director C. Shaler Smith moved that the Society appoint a committee to draft a law covering the points in the report and to include in addition "the necessary provisions to secure the inspection by experts of all questionable bridges now in existence." The motion stated that the law so drafted be recommended for adoption by the state legislatures, and that the members lend their support to this aim. At the same time, Clemens Herschel (President, 1916) urged appointing a committee "to draft a law requiring tests of finished bridges, before, and at stated times after, their opening for public travel."

Both proposals were put to letter ballot of the membership, as required at the time, and both were soundly defeated. The deciding argument appeared to be: "The Society, as a body, is not competent to set forth opinions on any special points of practice."

The Board did not always wait, however, for membership approval. The Committee on Tests of American Iron and Steel had been quite active since its beginning in 1872. Through its efforts, a bill calling for a national commission of experts to execute tests on structural materials and to "deduct useful tools therefrom" was introduced into Congress in 1882. When the legislation failed to pass, the Board of Direction moved unilaterally in 1883 to select a qualified panel "to prepare and promote such a programme of tests of structural material as to secure the best results possible from the Watertown Arsenal Experiments."

The number of bridge failures declined after 1900, but a most spectacular one was the collapse of the Tacoma Narrows Bridge ("Galloping Gertie") in 1940. Soon thereafter, President John P. Hogan was authorized to appoint an investigating committee, but the Society decided it would be better undertaken as a joint project with other organizations. ASCE appealed successfully to the federal Public Roads Administration to sponsor the project, and the Advisory Committee on Investigation of Long-Span Suspension Bridges was the result. This group developed valuable guidelines on the testing of and research for suspension bridge design.

The importance of regulating engineering practice in the public interest was highlighted by the May 1874 failure of the Mill River Dam, in Connecticut, which cost 143 lives. James B. Francis (President, 1881) headed an ASCE investigation panel. Extracts from the committee's report tell the story:

> The specifications ... were prepared by Mr. Lucius Fenn, Civil Engineer ... who, according to the evidence reported to have been given by him at the coroner's inquest, claims to have written then under the direction of the Directors,—"he acted only as the attorney of the company in drawing up the specifications." ...
>
> In the construction of the work by the contractors, it appears that there was no sufficient inspection, so peculiarly important in a work of this description, and during part of the time none at all, except by the Directors of the Company or their building committee, during their occasional visits. The remains of the dam indicate defects of workmanship of the grossest character ... it is obvious that this cannot be called an engineering work. No engineer, or person calling himself such, can be held responsible for either its design or execution.

The 1876 failure of the Lynde Brook Dam at Worcester, Massachusetts, was investigated by a special ASCE committee, chaired by Past-President Theodore G. Ellis. The three-man committee visited the scene and reviewed the available data, noting that it "was obliged to procure the information desired from other sources" when it was denied access to the plans from which the dam was constructed. The report of the committee contained the following conclusion:

> There is no doubt as to what was the immediate cause of the failure of this dam. There was evidently a stratum of porous material lying under the upper gatehouse and the upper end of the pipe vault, partaking of the nature of a quicksand, which should have been removed, and greater precautions taken to prevent the access of water from the reservoir, than appears to have been the case in the foundations above described.

The City of Worcester engaged Past-President William J. McAlpine as a member of a panel to investigate the failure. He stated tersely that he could not "assent to the conclusions of the report" and that he hoped to be able to report later on his own conclusions. He did not, however, take issue with the committee report in his recommendations for repairing the dam.

The devastating Johnstown Flood in 1889 resulted in a special working party "to visit the scene of the disaster and report to the Society the cause of the calamity." Six months later, the Committee to Investigate the Failure of the South Fork Dam was ready to report but stated that

> the Committee have considered that the presenting of the report at this time would not be a proper thing. Lawsuits involving claims for very heavy damages have been instituted by many of the sufferers ... and these suits are now pending ... and the Committee ... thought that to give publicity at this time to their conclusions might prejudice the case and ... affect ... the trials. The Committee have, therefore, agreed to recommend ... that the report ... should be placed in the custody of the Chairman of the Committee, Mr. James B. Francis, and kept there to be called for at any time after the issue of the trials in court has been decided.

Calling for presentation of the report at the 1891 Annual Convention, the Secretary stated: "The time for personal liability is now past and the principal suit has been decided that it was absolutely impossible to have foreseen this disaster; in other words, that it was the act of God." Following are extracts from the report:

> There can be no question that such a rainfall had not taken place since the construction of the dam. ... The spillway, however, had not a sufficient discharging capacity; contrary to the original specifications of Mr. W.E. Morris, requiring a width of overflow of 150 feet and a depth of 10 feet below crest, which would have been a sufficient size for the flood in the present case—it had only an effective width of 70 feet, and a depth of about 8 feet; the accumulated water rose to such a height as to overflow the crest of the dam and caused it to collapse by washing it down from the top. ...
>
> There are today in existence many such dams which are not better, or even as well provided with wasting channels as was the Cone-

> maugh Dam, and which would be destroyed if placed under similar conditions. The fate of the latter shows that, however remote the chances for an excessive flood may be, the only consistent policy when human lives … are at stake, is to provide wasting channels of sufficient proportion and to build the embankment of ample height.

Widespread incidence of "floods of exceptional magnitude" in the spring of 1913 resulted in the Society's authorizing a Committee on Floods and Flood Prevention. The committee pointed out that attention must be given the fundamental data, and the report encouraged state and federal agencies to collect such information systematically.

In 1922, the American Red Cross sought and received ASCE's cooperation in its flood prevention efforts. The Board of Direction acted in 1928 following flooding in the Mississippi River Valley, urging President Hoover to create a professional Board of Review on the Mississippi River flood control program, and to seek Congressional authorization for funds needed for surveys and compilation of data. A strong committee conveyed the recommendations to the President and Congress.

The need for flood information resulted in a new ASCE Committee on Flood Protection Data in 1923, and a Committee on Flood Control in 1936. Both committees were discharged after reporting in 1940, when their functions were relegated to the Hydraulics Division for "continuing study of flood protection data and determination of principles of practical application of such data to the design and operation of flood control works."

The urbanization of America was occurring so rapidly in the 1880s that provision of safe, potable municipal water supplies became a real problem. ASCE found an opportunity for public service in this situation. The following resolution was adopted at the 1889 Annual Convention:

> It is a well known fact that many cities and towns on the Atlantic Coast have suffered very greatly from impurities in their water supplies … and that no adequate remedy meeting all conditions has been found. …
>
> These impurities are often due to natural causes which have not been … investigated on account of the difficulty of centralizing the individual efforts of all parties engaged in such investigations,
>
> *Resolved*, That a Committee … ascertain the best means of concentrating all available information … to secure useful results and to report what further action should be taken.

In January 1890, the Committee on Impurities in Domestic Water Supplies recommended that a complete bibliography of physical, chemical, biological, and public health information be developed and that such references be assembled; further, that a system be set up to gather data from boards of health, waterworks managers, federal bureaus, and researchers throughout the country. Its report ended with the statement:

> If the engineering profession desires to retain the right to act as the final arbiter in the selection of sources of water supply, it must ... take the lead in ... scientific inquiry into the conditions affecting the purity of stored water and the subtle differences that separate waters of good quality from others, equally good in appearance and chemical composition, that do not stand this test.

Chairman Alphonse Fteley added that the American Water Works Association (AWWA) had recently established a similar committee, which was already gathering information. The Board authorized him to exercise discretion with regard to joint or independent action on the part of ASCE. An effective joint activity did not develop. The independent studies of the ASCE working party concluded that it was not feasible to attempt to assemble and collate the necessary data without substantial outside funding. The correspondence and circulars the committee distributed to boards of health and various water supply agencies emphasized the need for higher standards of domestic water quality. In 1893, the committee concluded its efforts to consolidate data. However, their work surely contributed to efforts of the American Academy for the Advancement of Science, the American Public Health Association, and AWWA to create the nation's first U.S. Public Health Service (PHS) Drinking Water Standards in 1914.

Catastrophic earthquakes have plagued the world from its beginning. The civil engineer considers seismic events in designing and building structures that can withstand the damage and hazard to human life.

When the San Francisco earthquake occurred in April 1906, ASCE appropriated $1,000 for the prompt relief of the city's engineers. More important, the San Francisco Association of Members offered to serve the stricken city in any way possible. The mayor appointed seven ASCE members to his Committee of Forty on Reconstruction. The Association also created six technical subcommittees to study the effects of the earthquake on structures. Their comprehensive report was published in March 1907 and provided invaluable guidance to designers and public authorities on earthquake and fire-resistant construction.

Again in 1923, following the major Japanese earthquake, the Society quickly communicated with its members in that country, and it authorized a committee of experts to conduct an "investigation of the effects of earthquakes on structures in Japan and elsewhere." The panel included engineers from both sides of the Pacific, and a voluminous report was submitted to the Society in 1929. Financial stress due to the Depression prevented publication, but it was available in the Engineering Societies Library as a reference on design of structures subject to seismic forces. By this time the Structural Division was well established, and members have analyzed seismic and other dynamic forces through the years, notably the investigations of the effects of the Florida hurricane of 1926 and of the Santa Rosa, California, earthquake of 1971. The Long Beach, California, earthquake of 1936 prompted the Los

Angeles Section to join the Structural Engineers Association of California and the Los Angeles Engineering Council in a study of damaged structures. Their joint report included recommendations on earthquake-resistant design.

Since 1900, ASCE has recognized the possible legal implications incident to liability suits and damage claims in conducting these investigations. Society members frequently serve on investigative commissions, but in recent years the Society's public safety efforts have been the work of the Technical Divisions. These studies cover such topics as flood control, structural failures, fire protection, public health, highway safety, earthquakes, hurricanes and tornadoes, landslides, and similar hazards to the public safety.

Public Works and Policy Guidance

Civil engineering practice is heavily involved public works, and ASCE has followed policy decision making in the fields of transportation, conservation, water resources, environmental quality, and public works administration, planning, financing, and management.

ASCE was still in its infancy, in 1870, when it authorized $250 for an ad hoc committee to be "charged with collecting such documents and information in relation to inter-oceanic communication between the waters of the Atlantic and Pacific as they can obtain." President Craven himself chaired the committee. Just a month later, Society adopted the following resolution:

> The relations of the Isthmus of Tehuantepec to the United States are so peculiar and so different from those of any other route for transit, that it is highly important that the patronage and influence of the government ... should not be committed to any ship canal or transit enterprise until the practicality ... of a canal across the isthmus has been determined by a survey.

The resolution was transmitted to both Houses of Congress.

During the next decade, the Society published many papers on the feasibility of an Isthmian canal at various locations. In February 1880, the Society sponsored a public meeting in the theater of the New York City Union League Club, at which Ferdinand de Lesseps led a general discussion on "Inter-Oceanic Canal Projects." According to *Frank Leslie's Illustrated Newspaper*, de Lesseps "delivered an address which for style, charm, interest and brilliancy should serve as a model for scientific lectures in general, and those in this country in particular."

An 1873 survey of national problems by the Committee on Library offered public policy guidance as well as technological study. Of 37 subjects relating to the practice of engineering, the majority dealt with transportation. Three items studied are still a major concern a hundred years later: rapid transit for large cities; pollution of rivers by sewage; and the production of mineral oils.

In September 1874, the Society approved the following resolution, which is still relevant today:

> *Resolved*, That a committee ... investigate the necessary conditions of success, and recommend plans for: The best means of rapid transit for passengers, and the best and cheapest methods of delivering, storing and distributing goods and freight, in and about the city of New York, with instructions to examine plans, and to receive suggestions such as parties interested in the matter may choose to offer, and to report on or before the first day of December, 1874.

Only three months were allowed for the resolution's fulfillment. Although not a policy determination, the 80-page report published a year later by this committee provided urban transportation policy guidance for larger cities for some years. Chaired by Octave Chanute, the committee introduced a number of new ideas, such as the suggestion by Richard P. Morgan Jr. that an elevated rapid transit trackway be constructed over the center of the street.

Strangely, there appears to have been no official ASCE involvement with railroad transportation policy, despite the explosive development of railroads from 1850 to 1900. However, prominent members critically discussed national railroad policy at Society meetings and publications. The contributions in this area by Albert Fink, Arthur M. Wellington, Octave Chanute, Ashbel Welch, John B. Jervis, Martin Coryell, and Albert Sears were especially noteworthy.

After 1950, the Society's involvement with national highway and urban transportation policy increased. A special task committee provided input to the National Highway Bill in 1956 and 1959; a report, "Principles of Sound Transportation Policy," was adopted in 1963; ASCE submitted recommendations on the federal Highway Aid Program (1970), highway safety (1971), rapid transit service to airports (1971), the Highway Trust Fund concept (1972), and a proposed reorganization of federal transportation agencies (1972).

Members protested the Society's 1973 opposition to using funds from the federal Highway Trust Fund to support urban mass transit studies. Many members argued that this position, which originated in the Transportation Policy Committee, did not represent the views of a majority of the membership. There was demand for membership ballot on policy issues, as had been done in the early years of the Society.

ASCE's concern for conservation of national resources is long-standing. The following resolution was among the incentives that led President Theodore Roosevelt to call a White House Conference of Governors in May 1908, to which he invited the presidents of the four Founder Societies:

> *Whereas* the timber resources of this country are being rapidly diminished owing to unscientific methods of forestry, to the prevalence of forest fires and a wasteful use of lumber ... which may result more-

> over in the diminution of the natural storage capacity of our streams, and increasing irregularity in the flow, and ... impairment of the value of our water powers:
>
> *Resolved, That* ... every endeavor should be made to further the introduction of principles of scientific forestry and the creation, preservation of National and State forest preserves ... and the Board of Direction approves and urges the passage by Congress of a bill providing for national forest preserves in the Appalachian and White Mountains.

The post-conference declaration confirmed the need for conservation, it urged policies and legislation to ensure the best use of forest, water, and mineral resources, and it recommended that a national Commission on Conservation of Natural Resources be established. President Theodore Roosevelt was not only receptive; he invited ASCE President Charles MacDonald to serve on his presidentially appointed national commission. After the White House Conference, the Founder Societies jointly sponsored a series of public meetings on conservation of natural resources during the years 1908–9.

In 1923–24, ASCE supported the Weeks Law, which provided for federal appropriations to purchase land for public use in the conservation of forest and water resources. Again, in 1927, the Society endorsed legislation fostering the purchase of land for forest development.

Although ASCE had a Committee on Impurities of Domestic Water Supply from 1890 to 1893, the Society was more interested in improving water potability than in the broader aspects of water resources policy. However, a broad view was implicit in a resolution adopted in May 1913:

> *Resolved,* That the Board of Direction ... appoint a Special Committee to investigate the advisability of drafting a National Water Law applicable to all navigable interstate and other waters within the jurisdiction of the United States, and embracing all uses of water, and that such committee be directed to prepare a preliminary draft ... if ... it appears advisable.

Apparently it was not found feasible, because the committee was replaced in 1917 by the Committee on Regulation of Water Rights, which was concerned more with state control than with water as a national resource. The committee functioned until 1920.

Stream pollution as a national problem came to official attention in 1922 when a member, Charles Haydock, urged the Society to study pollution of streams by industry and the need for legislation in this regard. The Board of Direction responded by taking the following action:

> *Resolved:* That the United States Government be requested to undertake, through the Department of Commerce, a complete investigation

> of the cause, extent and effect of pollution of waters by industries, that methods of mitigating such evils be investigated, and that existing legislation be reviewed to determine what if any legislation is required.

This seed, unfortunately, fell upon barren soil. The only federal official to respond was the Secretary of War, who merely referred to the modest appropriation provided to the Public Health Service for studies of stream pollution.

One of the most significant contributions ever made by the engineering profession to national water management was a report, "Principles of a Sound National Water Policy," produced in 1951 by the EJC National Water Policy Panel, and the 1957 "Restatement" of that report. The EJC Panel was largely composed of ASCE members.

By 1960, ASCE had become directly involved in water resource matters. A top-level Committee on National Water Policy, set up in 1961, devoted its full attention to that purpose. In 1963, the committee reviewed 19 Congressional bills of major concern to the Society. The committee maintained effective communication with all federal water agencies, such as the National Water Commission and the Water Resources Council.

Although water pollution legislation was the major issue throughout the 1960s, ASCE's Water Policy Committee also considered other problems, including water resource planning, watershed protection, flood prevention, flood insurance, wild and scenic rivers, estuarine preservation, reclamation, and coastal management. In 1972, the committee prepared an analysis of the National Water Policy Commission's report, "Principles and Standards for Planning Water and Related Land Resources."

On the international level, ASCE endorsed a 1922 recommendation of the International Joint Commission of the United States and Canada that a special technical board investigate and report on proposed improvement of the St. Lawrence River as a waterway. ASCE supported legislation to authorize a commission "to give further study to the essential facts with reference to a St. Lawrence Waterway, to guide the action of Congress with respect to its construction" and appealed to the appropriate committees of Congress. The Society also urged other Founder Societies and the Federated American Engineering Societies to assist in this effort.

Some public policy concerns of the Society were not limited to the engineering profession. At the 1924 Annual Meeting, an ASCE resolution supported federal legislation that would reduce personal income taxes. The action was taken unanimously and without discussion!

Another significant ASCE public service effort occurred during the Great Depression. A group of 16 members of the Society, headed by John P. Hogan (President, 1940) drafted a memorandum, "A Normal Program for Public Works Construction to Stimulate Trade Recovery and Revive Employment," which proposed creating a federal credit corporation. The principle was adopted on May 9, 1932, by this resolution:

Resolved, that the American Society of Civil Engineers, through its Executive Committee,

1. Approves in principle a normal program of public works construction as the most effective immediate means of increasing purchasing power, stimulating recovery and reviving employment.

2. Urges on the Congress of the United States the enactment of the necessary legislation to extend Federal credit facilities to solvent states, counties and municipalities to enable them to carry out their normal programs of necessary and productive public works.

A special committee presented the recommendations to the President in person. President Hoover invited the Society to assist in drafting legislation, and also asked ASCE for data on the nature and cost of public works construction being deferred for economic reasons. Through a survey of Local Sections, the Society identified a backlog of more than $3 billion in such delayed construction, and it offered these results to the President and to Congress.

On July 16, 1932, both Houses of Congress passed the Emergency Relief and Construction Act of 1932, and President Hoover signed it on July 21. The bill, which included the financing principle espoused by ASCE, resulted in an updated Reconstruction Finance Corporation.

The Society's Committee on Public Works also supported the National Industrial Recovery Act of 1933, which authorized the Public Works Administration. These laws not only improved the national welfare, they resulted in employment for more than an estimated 40,000 engineers.

In addition to Colonel Hogan, the Committee on Public Works included former and future ASCE Presidents Harrison P. Eddy, Malcolm Pirnie, and Alonzo J. Hammond, and future Vice President Joseph Jacobs.

In 1928, the National Committee on Calendar Simplification asked the Society for support in this effort. Such calendar reform would be a long and arduous undertaking, and the Board of Direction approved the proposal to urge the Secretary of State to involve the United States in the international deliberations on the question then under way.

Five years later, in June 1933, the Board adopted the following resolution:

Resolved that the American Society of Civil Engineers, in view of its resolution adopted in 1929 in favor of improving the calendar, notes with satisfaction that in 1931 an international conference of the League of Nations (Fourth General Conference on Communications and Transit), in which the United States Government participated, officially considered this question, and that the conference … recognized that the simplification of the calendar was certainly desirable, and gave to the governments a survey of the question for their future decision.

Resolved that the Society … express to the Secretary of State the hope that our government will indicate to the League of Nations a

> desire to have this question again taken up at this conference and to again participate in the discussion.
>
> *Resolved* that the Society is of the opinion that a reform based on a division of the year into 13 equal months would best adjust the calendar to modern conditions.

This policy of the Society still prevailed in 1974.

The four Founder Societies combined in 1946 when their presidents communicated a joint statement to the Senate Committee on Atomic Energy urging that responsibility for nuclear energy development be placed in a civilian commission. Such a body was created in the Atomic Energy Commission (AEC), and in 1952 ASCE formed a Committee on Atomic Energy to study and recommend policies and actions relevant to the interests of the Society regarding nuclear energy. For several years, ASCE communicated with AEC in reviewing both manpower and contract negotiation issues and technical issues of power plant design and environmental impact.

By the early 1970s, ASCE was closely attuned to federal policy determination over a wide spectrum. Two timely ventures in 1973 were the Land Use Policy Committee and the National Energy Policy Committee. The land-use policy action anticipated federal land-use policy legislation, and urged grassroots implementation at the state and local levels. The national energy policy panel sought to establish a database and to devise a coordinated energy conservation program to manage short-term dislocations in the energy supply. Other legislative actions in 1974 included weather modification, negotiation of professional service contracts, environmental quality education, housing, transition to the International System of units of measurement, and ocean resource development.

Relationships with Government Agencies

Many civil engineers are employed by agencies of government at all levels, and many government bureaus, particularly federal ones, provide services closely related to the professional practice of civil engineering. As a result, ASCE is keenly interested in the laws, regulations, and procedures that dictate government operations. Only a few Society actions are summarized here.

In 1878 a "memorial" to Congress gave ASCE endorsement to the extension of the national system of triangulation by the U.S. Coast and Geodetic Survey into all jurisdictions where surveys were authorized by the state level. Triangulation was especially important for accurate and convenient location of the roads, railroads, waterways, and other public works construction. Many later actions dealt with surveying services—triangulation, topographic, seismic, oceanographic, and so on—and reached a peak during the Depression years as a source of productive employment for engineers. In 1934, the Society opposed transferring the Survey from the Department of Commerce to the Navy Department. The transfer did not occur.

Operations of the federal Bureau of Reclamation are hard-core interests of ASCE, and many engineers employed in the Bureau are members of the Society. The reaction to the politically inspired dismissal of Arthur P. Davis (President, 1920) as Director of the Bureau in 1923 was strong (see Chapter 5). The purge by Interior Secretary Hubert Work resulted in the dismissal of 12 staff engineers and 15 consultants. Secretary Work was forced to set up a special Fact-Finding Commission in an effort to justify his actions. Noting that "no evidence has been provided sufficient to warrant so phenomenal an overturning in the Reclamation Service," the Society made clear its unhappiness with the whole situation.

The Society took note of certain expressed aims of President Warren G. Harding, and later President Calvin Coolidge, to streamline the federal bureaucracy, and it embarked upon a similar undertaking in 1923:

> *Resolved:* That the Board of Direction of the American Society of Civil Engineers endorses and commends the recommendations of the President of the United States and his Cabinet, that the military and nonmilitary engineering activities of the Government be separated, and that the design, construction and maintenance of nonmilitary public works be assembled as far as practicable in one department, under one head, and that only those activities closely related be included in that department. We also commend the effort to apply similar principles to all the departments, and ... believe such action will tend to eliminate duplication, to coordinate public activities, and in many ways to promote economy and efficiency in the public service.
>
> *Resolved:* That the President of this Society be empowered to appoint a committee of five members of this Society of which he shall be the Chairman, to present the above resolution to the President of the United States, and to appropriate officials of the Congress, and of the Executive Departments, and to take such other action as it deems wise.

The Society appropriated $1,000 to cover expenses arising from the resolution.

Society President Charles F. Loweth and his Committee on Federal Reorganization presented in person to President Coolidge, on September 30, 1923, the documentation setting forth the position of ASCE. To bring about professionwide support, the committee also enlisted the sponsorship by Federated American Engineering Societies of a Conference on Public Works in Washington, in January 1924. About 60 engineering and architectural organizations participated and reacted favorably.

Naturally, there was outspoken opposition by the affected federal bureaus, highlighted by an appearance before the Board of Direction by General Lansing H. Beach, Chief of the U.S. Army Corps of Engineers. A bitter exchange took place between the Society and General Beach, following the hearings before the Congressional joint Committee on Federal Reorganization. He and other ASCE members in the Corps considered the position of the Society to be contrary to their interests.

The reorganization legislation was not enacted. Nevertheless, the Society was consistent when it responded to the Wyant Public Works Bill, in 1928, by adopting the principle "that reorganization and concentration of the engineering functions of the Federal Government is desirable and advisable and the Board of Direction favors such legislation," and again, in 1937, when it supported the recommendation of President Franklin D. Roosevelt that a federal Department of Public Works be established. But, as before, Congress did not agree.

ASCE was forced to take sides in 1928 when there was a movement to put a proposed National Hydraulic Laboratory under the direction of the Corps of Engineers after the Senate passed a bill authorizing the laboratory in the National Bureau of Standards (NBS). The Society supported operation of the laboratory in the NBS, but the legislation did not pass. When Congress appropriated funds in 1931 for the NBS laboratory, ASCE made an immediate offer to advise and assist in its design to ensure that it would be "fully adapted to the study of the problems which the tremendous natural resources and great industrial progress of this country must ... bring to it for solution." A blue ribbon committee was set up to implement the offer.

Through the initiative of its Sanitary Engineering Division, ASCE pressed for years (1920–24) for legislation authorizing sanitary engineers to be commissioned officers in the PHS. This would enable these engineers to collaborate with other professionals, particularly medical personnel.

A 1924 "Bill to Promote the Efficiency of the Public Health Service" failed to gain the support of the Budget Director. Other similar measures also failed, for various reasons. It was not until 1943, 19 years after the original proposal, that a suitable PHS reorganization measure became law. Aggressive action by the Society through its Washington staff was responsible for two important amendments favoring the status of sanitary engineers.

A regrettable sequel came in 1971 when the PHS lost most of its responsibility in public health engineering to the new Environmental Protection Agency. Commissioned status for sanitary engineers was discontinued at that time, despite the Society's protests.

ASCE's endorsement helped the U.S. Geological Survey in 1925 by bringing about legislation authorizing a continuing inventory of national water resources. A strong resolution urged appropriation of the necessary funds without a time limit on the work, but it recommended that the program not interfere with the regulation by the states of any streams wholly within their boundaries.

The Society often appealed to the President of the United States. In 1935, the Public Works Administration (PWA) sharply amended its project review procedures with the result that several thousand projects—prepared in compliance with previous requirements—were summarily rejected. The following telegram was among those that brought about restoration of the original PWA operation:

> The Honorable Franklin D. Roosevelt
> Hyde Park, New York
>
> The wholesale rejection of PWA projects prepared in accordance with the recommended and approved procedures breaks faith with applicants who ... have advanced time and money in the preparation of plans for sound and useful projects (stop). Stoppage of proposed worthwhile PWA projects denies opportunity to engineers, architects, skilled workers, and business organizations in the construction industry to continue along lines of their normal training and operations thus throwing more into the ranks of the unemployed (stop). American Society of Civil Engineers strongly endorses the basic principles of PWA procedure (stop). You are urged to correct the injustice of the present situation.
>
> George T. Seabury,
> Secretary, American Society of Civil Engineers

A special ASCE Committee on Atomic Energy was formed in 1953 to cooperate with the AEC in the advancement of nuclear technology. The AEC welcomed the Society's interest.

Many requests for improved or expanded service in technical agencies of the government originate in ASCE's Technical Divisions. Typical was the work done by the Irrigation Division in the mid-1920s, aimed at enhancing cooperation between state agencies and federal bureaus in the Department of the Interior and the Department of Agriculture that dealt with irrigation. A report published in 1927 offered several useful recommendations. Another example was the strong plea in 1936 for an expanded program in the Weather Bureau to provide basic data required for development of water and agricultural resources, aviation, and public works construction. The resolution drafted by the Committee on Meteorological Data and adopted by the Board was specific in its recommendations, and it contributed to bringing about an increased appropriation.

Communication with federal agencies is a continuing day-to-day part of the ASCE headquarters operation. The relationship extends to all bureaus that have any responsibility with construction or with engineers. This effort ranges from cooperation in technical affairs to personnel problems, to contract negotiations, to public works administrative procedures, and beyond. The complexity of such issues will likely increase, with the trend toward the expansion of the federal bureaucracy.

Wartime Service

In February 1917, before the entry of the United States into World War I, ASCE President George H. Pegram joined the leaders of the other Founder Societies in pledging to President Woodrow Wilson their united support in the stand for freedom and safety of the seas, and offering to assist "organization of engineers for service to our Country in case of War."

Only three months later, the United States was embroiled in the war. The Board of Direction expressed itself on the danger of relying on volunteer armies and navies by resolving: "That Congress ... pass at once a bill providing for universal military training and service, which we hold to be the only proper, democratic and efficient way for creating the public defense." At the same time, ASCE urged that all possible consideration be given to students in recognized engineering schools to enable them to continue their education.

The War Revenue Act of 1917 created a problem for ASCE when the professions were interpreted to be included in the trade or business categories and thus subject to an 8% excess profit tax. Two local associations openly opposed the effort of the Society to obtain exemption of professionals from the tax, holding the action to be unpatriotic. From the record, however, the Society appeared to be well justified in its objection to an unfair and improper tax on professional fees. About 15% of the membership of the Society was in some branch of the military or related wartime service during the period 1917–19.

At the 1919 Annual Meeting, R.S. Buck proposed that the Society urge upon all public works authorities

> that public works should be carried forward to the fullest possible extent consistent with sound judgment, not only for fundamental economic reasons, but for humanitarian reasons, to furnish employment for all who can properly claim employment, especially returning soldiers.

A post–World War I proposal (1923), upon which the Society acted negatively, is interesting because of the widespread antiwar sentiment generated by the Vietnam War in the late 1960s. The following resolution offered at the 1923 Annual Meeting was promptly tabled:

> *Whereas*, Many engineers, including members of this Society, actively supported the recent World War, which was entered by the United States with the avowed purpose of preserving civilization in a war to end war and to make the world safe for democracy; and
>
> *Whereas*, It is now evident that the World War did not end war, and that present conditions in Europe may soon lead to another war; and
>
> *Whereas*, The Engineering Profession is devoted to the service of humanity in directing the great sources of power in Nature for the benefit of the human race, not for its destruction; therefore

> *Be It Resolved,* That the members of the American Society of Civil Engineers in Annual Meeting assembled this 17th day of January, 1923, hereby declare their opposition to human warfare and their refusal to support war for any purpose or at any time; and
>
> *Be It Further Resolved,* That the Board of Direction be requested to take such action as will aid in the establishment of an international economic commission for the purpose of regulating international commerce and directing the development of the resources of the world for the benefit of all mankind.

However, when Secretary of War John W. Weeks sought the assistance of the Society regarding War Department construction operations, the Society's response was quite different. Five representatives were promptly appointed to attend a conference in January 1925, aimed toward the cooperation of the entire construction industry with the War Department.

In 1922, the Attorney General indicted several ASCE members for alleged improprieties in World War I military construction. In response, a strong ASCE resolution demanded prompt action so that the allegations could be sustained or disproved. The courts found no criminal conspiracy had been involved in the cost-plus construction contracts. Because of the war and the need for speedy construction, these were used instead of the customary competitive bids. The accused members were completely vindicated.

In July 1940, the Board created a Committee on Civilian Protection to be concerned with "all matters relating to civilian protection; the safeguarding of life and of civilian activities in general and the protection of public utilities." As a result, 62 local civilian defense committees were formed.

In July 1941, the Board and the membership adopted a resolution endorsing President Roosevelt's program and efforts in the national defense. These actions were reiterated in January 1942 just after Pearl Harbor. By this time, the new Washington Field Office was well established, and it acted forcefully in the critical areas of military and civilian manpower, wartime construction controls, and planning for postwar construction.

In 1943, the Board of Direction became concerned as to whether engineers were fully utilized in the armed services. The following resolution went to the War Manpower Commission: "The Board urges (1) that a procurement board for engineers similar to that for procurement of medical officers be established; and (2) that it be conducted and operated in like manner to that for medical officers."

The Construction Division initiated a Committee on Postwar Planning (later Postwar Construction) in 1943 to study the conversion to peacetime industrial and public works activities. The Board of Direction approved the report the same year, and it was used to draft the public works bill submitted to Congress by ASCE President Ezra B. Whitman in January 1944. The bill proposed that all federal postwar construction be placed under a single agency and that planning funds for federal construction grants be provided. Congress did not accept the idea of a single public works agency, but other ASCE recommendations became part of the legislation. The Committee was discharged in 1945.

As the end of the war appeared to be approaching, President Malcolm Pirnie joined with the presidents of AIME, ASME, AIEE, and the American Institute of Chemical Engineers in offering the assistance of the engineering profession in solving problems related to the demilitarization of Germany and Japan. Programs for postwar industrial control in the two countries were submitted to the Secretary of State in October 1944. As a result, Secretary Stettinius invited the Engineers Joint Conference Committee to pursue its studies in detail, and to develop its program with his office and the Foreign Economic Administration. This commitment was carried out by the Engineers Joint Conference Committee, which evolved into the Engineers Joint Council in 1945.

A 1943 membership questionnaire survey disclosed that more than 5,000 members—about 27% of the total—were then in the armed services. About 93% were commissioned officers, with approximately 35% in the Navy and 60% in the Army. The survey also showed that 40% to 50% of the membership was engaged in wartime construction, industry, and government service.

Any prior differences between the Army Corps of Engineers and civilian groups in the Society were forgotten now. Every assistance was given the Corps in its recruitment effort, and many future leaders of the Society served in its ranks. Similarly, in the Navy, the Air Corps, the Sanitary Corps, and other branches of the military members of the Society were generously represented. The Construction Battalion (Sea Bees) of the Navy had an outstanding wartime record, with ASCE members filling many leadership posts.

The Board of Direction—deeply conscious of the need to preserve the national defense in the age of nuclear weapons—resolved in October 1945: "That the establishment by the Congress of the United States of a system of universal military training be urged, to effect adequate National Defense and to promote the security and well being of the nation in time of peace and, if need be, to provide a maximum degree of protection and security in time of war." The Secretary of War and many members of Congress acknowledged the resolution.

Since 1962, the Society has been represented by its Executive Director on an advisory committee to the Secretary of Defense on "The Design and Construction of Public Fallout Shelters." In 1973, the Civil Defense Preparedness Agency of the Department of Defense agreed to work with ASCE to develop a training program for engineers and architects in the causes of building collapse and other structural failures.

Public Appointments

ASCE is often asked to nominate candidates for technical or other public commissions, or to provide special professional services. In 1875, the President of the United States asked the Society to name representatives to the U.S.

Commission on the Improvement of the Mouth of the Mississippi, and the Board complied. Similarly, when the Secretary of War asked ASCE to nominate three members of the U.S. Commission to Test Iron, Steel, and Other Metals, the Society did so. However, when the City of Providence asked ASCE to examine and evaluate a proposed sewerage plan for that city, the Society responded: "The Society as such cannot undertake to nominate committees to serve private interests, but the Board of Direction [may] be requested to transmit to the mayor the names of a number of experts from whom he may select gentlemen to serve on the committee referred."

The distinction was sometimes unclear among appointments to public commissions and advisory bodies, appointments to public office, and requests to perform consulting services for a fee. At the 1875 Annual Convention, the Board decided: "It is inexpedient for the officers of the Society to take action on applications for members to perform professional services."

This ruling was not considered mandatory. ASCE furnished lists in 1879 when the cities of Holyoke, Massachusetts, and Milwaukee sought consultants, and again in 1882–83 when the City of Philadelphia set up expert panels to study water supply and street improvements. ASCE attempted to set policy on these issues again in 1905, stating

> that the nomination by the American Society of Civil Engineers, or by its officers, of persons to serve on technical commissions, or to render special professional services, does not come within the purposes for which the Society was organized, and exists, and that the making of such nominations is not advisable.

But again the Society officers did not construe this language as prohibitive, and usually offered nominations when requested, typically lists of names. The Society rarely endorsed or evaluated single candidates, unless the appointing authority asked them to do so. No reason is recorded for a 1917 resolution regarding policy on appointments to public office: "No officer of the Society shall officially recommend anyone for any office or position."

This move was reconsidered by the Board of Direction in 1922, and tabled. Because ASCE never rescinded the resolution, it would still seem to prevail in 1974 as Society policy.

The 1923 effort to do away with engineering leadership of the Bureau of Reclamation must have been partly responsible for the following 1926 declaration concerning the qualifications of appointees to certain public positions:

> *Therefore Be It Resolved* by the Board of Direction of [ASCE], that public positions and offices whose incumbents are charged with duties requiring engineering or other technical training should be filled with properly qualified persons, and that whenever positions or offices are to be filled concerned with … engineering matters, every … effort should be made to … [fill them with] competent members of the Engineering Profession.

In 1953, the Board of Direction voted on a statement that affirmed the earlier statement and formalized the manner in which the Society would nominate candidates for public positions: "The Society continues to support the principle that engineers be appointed to fill engineering positions in government. ... When such slates are submitted, it should be stated that they are not to be considered all-inclusive."

Regrettably, the tendency to appointing unqualified political activists to posts requiring engineering judgment and expertise appears to be growing. The Bureau of Reclamation has been particularly susceptible, beginning with the ouster of ASCE Past-President Arthur P. Davis in 1923 (see Chapter 5). A similar situation in the Bureau drew fire from the Society again in 1934, and some later Bureau of Reclamation appointments were highly controversial.

In 1927, Past-President John F. Stevens sought to appoint competent civil engineers to the Interstate Commerce Commission (ICC). When 1941 legislation proposed limiting practice before the ICC to members of the bar, the Society contacted the Judiciary Committees of both Houses of Congress, stating that "the work of the Commission will not be improved by excluding non-lawyers from practice before it."

Since 1955, ASCE has encouraged the appointment of qualified civil engineers to posts in many federal agencies, such as the St. Lawrence Seaway Advisory Board, Department of Housing and Urban Development, Department of Transportation, Department of the Interior, and the several interstate boundary commissions. The extensive reorganizations triggered by the environmental-quality movement in the late 1960s demanded a great deal of Society attention, because of claims by conservation extremists that engineers were not sufficiently aware of the ecological impact of their works. Lawyers and other nonengineers were often appointed to policy-making posts in agencies oriented to engineering output. Their unrealistic objectives did not always advance the cause of environmental protection.

ASCE has tried to help public agencies and the private sector to identify qualified consultants. Since 1955, such requests for the names of consultants invariably have been handled by referral to the Professional Directory featured in *Civil Engineering*. The Society furnished shorter lists in some special cases.

ASCE adopted a position paper, "Policy Regarding Government Agencies Employing Professional Engineers," in 1948 (updated in 1959). The paper, still in force in 1974, reinforced the importance of qualified engineer-administrators, and it offered guidance for the relationship between engineers employed in government and those who serve public agencies as consultants.

Public Relations and Information Services

In 1868, William J. McAlpine, ASCE's third President, stated the greatest need of the Society: "To make its advantages more generally known to the profession and to the public, and thereby obtain for it a higher standing and more influence with both." However, except for the Society's participation in several international expositions, no real effort was made toward public education and recognition until the twentieth century was well under way.

W.M. Hoyt, M.ASCE, told the Board of Direction in May 1921 that he did not see "why a campaign of advertising cannot be undertaken to let the public know something of what the Society stands for and make the term, Engineer, mean something." At the same time, the 1919 Committee on Development proposed two areas for inclusion in the new Constitution and Bylaws adopted in 1921. These were "Publicity" and "Public Affairs." The committee's recommendations were never adopted.

In 1921, the revised Bylaws provided for a new Public Relations Committee:

> The Public Relations Committee shall consider and report to the Board of Direction upon … matters of public policy and professional relations … and shall call the attention of the Board … to such matters affecting the welfare of the Society, or its members, or the Engineering Profession, as in its opinion should receive consideration or action.

The committee served as a legislative review and advisory body for nine years, greatly assisting the Board's consideration of public policy legislation ranging from conservation of natural resources to government reorganization. In 1923, the Board authorized subcommittees in the Local Sections, and in 1924, instructed the committee to initiate collaboration with the other Founder Societies to mount joint public affairs action.

In 1930, the Board replaced the Public Relations Committee by two new bodies: a Committee on Legislation and a Committee on Public Education. ASCE's public relations effort grew with programs including radio broadcasts, syndicated articles for the popular press, slide presentations, film bibliographies, and public education committees of the Local Sections. *Civil Engineering* magazine was used as an external relations medium as well as the primary means of communication within the membership. An intersociety task force sought commercial publication for material too costly for individual societies to print on their own, but this effort was only nominally successful.

In 1938, the Committee on Public Education became the Committee on Public Information, but it was discontinued in 1941 during a major reorganization. Another extensive reorganization in 1971 recreated the Public Relations Committee under a new Administrative Division. The committee was charged to

> develop a public relations program … improving the public understanding of the civil engineer and increasing his leadership in national, regional and local affairs. The … program shall be carried out at the national level through an in-house professional staff … supplemented by consultants as appropriate.

In 1935, the Board authorized retaining outside experts to prepare press publicity materials. The first in-house Department of Public Relations began in 1936, led by a professional journalist.

The Centennial of Engineering in 1952 marked the revival of the broad program of public education envisioned in 1935 when resources were lacking. Emphasis shifted from newspaper publicity alone to such media as radio, television, special brochures, and films; to recognition of civil engineering projects and accomplishments; and to enhancement of references to civil engineers and their activities in books, encyclopedias, and the popular press.

By Their Works You Shall Know Them

In 1962, the Biblical paraphrase "By their works you shall know them" summed up the ASCE public relations effort. For the first time, a professional specialist in oral, visual, and graphic expression headed staff operations, with an assistant to handle routine publicity. The program was very much of the "soft sell" type, but effective. The public was reminded of the contributions of the civil engineer, but members were also given cause for pride in their profession.

Public relations undertakings began with the "Seven Civil Engineering Wonders" project of 1955, a spin-off of the Centennial celebration in 1952. Local Sections were encouraged to select and publicize outstanding civil engineering works in their areas. There was so much local interest that in 1954 the Board of Direction set up machinery for designation of the Seven Modern Civil Engineering Wonders of the United States, which were announced in 1955 as Chicago's Sewage Works, the Colorado River Aqueduct, the Empire State Building, the Grand Coulee Dam, the Hoover Dam, the Panama Canal, and the San Francisco–Oakland Bay Bridge. The response from the news media and the general public was far beyond all expectations. *Reader's Digest* carried an 8-page feature given worldwide circulation. *Time* magazine devoted generous space to the announcement, as did leading newspapers throughout the nation. Thousands of reprints were distributed, and there was extensive radio and television coverage.

As an afterthought, the Society decided to prepare bronze plaques for each project site. Each plaque was unveiled in a public ceremony, generating even more publicity about the profession and the Society. The cost of the Seven Wonders project was nominal; its benefits were inestimable.

The Society decided that the idea merited a continuing format and established the annual Outstanding Civil Engineering Achievement (OCEA) Award in 1959 to recognize projects demonstrating both engineering skills and contributions to engineering progress and humankind. The first award went to the Saint Lawrence Seaway in 1960. All nominated projects in the national OCEA yearly contest received public recognition and a number of Sections initiated local OCEA programs for their own geographical areas.

The Seven Wonders theme continued to attract notice, and in 1962, the *Rotarian* magazine carried the story to its 22 million readers. A few months later, *Reader's Digest* published an article, "Five Future Wonders of the World," based on documentation supplied by the ASCE staff. In 1966, the staff assisted the author of a book, *Wonders of the World.*

In 1965, the Board established a Committee on History and Heritage of American Civil Engineering as part of the public relations program. One of its first efforts was the National Historic Civil Engineering Landmark program, which recognized early projects that advanced the profession and contributed to national development. The first such landmark was the Wendell Bollman Truss Bridge at Savage, Maryland, which was marked with a bronze plaque unveiled with appropriate ceremony in 1966. By 1974, a total of 36 National Historic Civil Engineering Landmarks had been identified.

The earliest historic landmark plaque was mounted by the Connecticut Section in 1937 at the Quinnipiac River Bridge near New Haven in memory of Clemens Herschel (President, 1916). Another special memorial plaque was unveiled at Wethersfield, Connecticut, in 1970, commemorating Benjamin Wright, 1770–1842, as "The Father of American Engineering."

In 1970, ASCE joined with the National Park Service and Library of Congress to create the Historic American Engineering Record, to document important early projects. The National Geographic Society, the Society for the History of Technology, the American Historical Association, and the Smithsonian Institution all cooperated in this venture. The Smithsonian maintains a Civil Engineering Museum and the Archival File.

The History and Heritage Committee introduced in 1970 the first of a series of "mini-histories," entitled "The Civil Engineer: His Origins." Other issues on canals, bridges, railroads, and the like were planned. Also, in 1972 the committee produced its first *Biographical Dictionary of American Civil Engineers.*

The American Bicentennial Celebration in 1976 promised to offer a unique opportunity for recognizing civil engineering accomplishments, and the committee assumed leadership in the bicentennial committee of EJC. It also worked on launching an ambitious theater-scale audiovisual presentation that portrayed the role of the engineering profession in the past and future development of America.

The growth of television during the 1960s offered other opportunities. In 1963, a one-hour show focused on civil engineering aspects of the New York World's Fair. Collaboration and guidance were provided for the major network documentaries, "Essay on Bridges" in 1965 and "They Said It Couldn't Be Done" in 1970. A videotaped presentation on "Metropolitan Planning and Design" was aired by 78 stations.

The 1968 half-hour film, "The Invisible E—the Civil Engineer," was shown by 250 television stations to more than 8 million people in its first year. By 1973, it had been telecast 744 times to a total estimated audience of more than 14 million and shown more than 400 times at group meetings to audiences totaling more than 16,000. A youth-oriented, career guidance film,

"Beginnings," was released in 1974 for showing on television and to student groups. A series of one-minute noncommercial spot films for televising on public service time seems promising. The first of these, "Water," was released in 1971, and it was followed by similar films on structural engineering, " people movers," transportation and environmental engineering, and professionalism in the community. By 1973, five of these films had been telecast 6,700 times on 791 stations before 262 million viewers.

The public relations staff was also responsible for the production of career guidance materials after 1960. These included special brochures, sound slide presentations, and the film "A Certain Tuesday" for showing in schools. As of 1973, the latter 14-minute film had been shown almost 367 times to audiences of 14,000.

A "Public Relations Guide," first issued in 1958, is a detailed manual for Sections. At least since 1938, some form of public relations newsletter has been circulated to the Sections, under such titles as "Headliners," "PR Bulletin," "ASCE Information Bulletin," "The Section Leader," and "PR Newsbriefs." Some Sections have mounted outstanding efforts, with news conferences and publicity about newsworthy members and their accomplishments.

The Society assists authors of engineering books and career guidance manuals, as well as editors of encyclopedias and dictionaries. A 1957 effort to develop a cartoon strip featuring a civil engineer was unsuccessful. In 1966, however, an educational ASCE cartoon, captioned "In This World," was released and was syndicated to some 4,500 weekly newspapers. Two more such cartoon series, "Building a Better World" and "News of Ecology," were also released.

In 1958, the Boy Scouts of America asked ASCE for help in updating its Surveying Merit Badge. This resulted, after 10 years of persuasion, in a new engineering scouting award in 1969. Another one-time venture was the 1964 exhibit on "20th Century Engineering" at the Museum of Modern Art in New York City. The Society supported this display, and it also collaborated in exhibitions elsewhere.

The ASCE public relations program relies on Society members as the most effective resource for building public respect and goodwill for the profession. The corporate annual report issued since 1956 and the editorials by the Executive Secretary Director in *Civil Engineering* since 1962 are further efforts. Both were meant to inform the membership and to arouse interest, enthusiasm, and participation by the members.

When it was recreated in 1971, the Public Relations Committee was evaluated by a professional public relations firm. The consultants found the operation to be "thoroughly professional." The same consultants made a detailed proposal for an expanded program in 1972–73, recommending additional funds to strengthen ongoing activities, and some new coverage.

CHAPTER 8

Working with Other Engineering Societies

The engineering profession is highly heterogeneous, more so than medicine and law. In the latter professions, a basic education of at least seven years is common to all; specialization comes after that. Next, there is a formal internship in medicine and preprofessional training in law. In engineering, the "common core" of education is usually no more than two years, after which the student moves into an engineering specialty and then, possibly, into narrower specialization. There is no formal internship except in the cooperative plans offered by some schools.

Relatively few doctors and lawyers are employed in government and industry. At this writing, about 30% of civil engineers are principals or employees in private practice consulting firms and 40% are employed by government. Engineers other than civil engineers practice predominantly in industrial organizations, with well under 10% of all others in private practice.

The medical and legal professions have universal and well-defined licensing procedures. Engineers have very little in common in their education experience, and a wide spectrum of practice, covering such diverse specialty fields as agriculture, electronics, construction, mining, chemicals, mechanical, electrical, public works, and aerospace. Moreover, the licensing arrangement for engineers is largely voluntary and not restrictive in its requirements.

Because ASCE was the first and is the oldest national engineering society, it was only natural that the Society would be cautious about yielding or sharing its identity, stature, autonomy, and resources with other emerging professional organizations. The public political orientation of civil engineering practice fostered different interests and attitudes than those engineering specialties geared to the industrial commercial world. This polarization often caused disagreement in intersociety deliberations.

At the same time, ASCE did not try to become an island unto itself. In most cases, the Society participated readily in joint and group liaisons and sought intersociety collaboration.

The American Society of Civil Engineers and Architects intended originally to foster professional improvement and social intercourse "among men of practical science" and "the advancement of engineering in its several branches and of architecture."

The notice addressed to prospective Charter Members following the Society's founding in November 1852 envisioned a unified profession, with one organization serving "Civil, geological, mining, and mechanical engineers architects and other persons who, by profession, are interested in the advancement of science." There was a note of realism:

> It is anticipated that the union of the three branches of civil and mechanical engineering and architecture will be attended by the happiest results, not with a view to the fusion of the three professions into one; but as in our country, from necessity, a member of one profession is liable at times to be called upon to practice to a greater or less extent in the others, and as the line between them cannot be drawn with precision, it behooves each, if possible, to be grounded in the practice of the others; and the bond of union established by membership in the same Society ... will, it is hoped, do much to quiet the unworthy jealousies which have tended to diminish the usefulness of distinct societies formed heretofore by the several professions for their individual benefit.

This optimism proved to be wishful thinking. Only a handful of architects ever became members of the Society, largely because the American Institute of Architects (AIA) was organized in 1857 while ASCE was hibernating from 1855 to 1867. The Society deleted "architects" from its name in 1869. The American Institute of Mining Engineers (AIME) was organized in 1871 to identify the second branch of American engineering, followed by the American Society of Mechanical Engineers (ASME) in 1880, and the American Institute of Electrical Engineers (AIEE) in 1884. Some of the earliest engineering-related technical associations were also formed during this period, such as the American Iron and Steel Association (circa 1875) and the American Water Works Association (1881).

There were never significant percentages of mining, mechanical, and electrical engineers in the membership of the Society. Presumably, the major branches of engineering began to emerge as specialties during the Industrial Revolution after the Civil War. The real fragmentation of the profession began after 1900, with narrower fields of specialization.

Civil Engineering and Architecture

Civil engineering is more closely related to architecture than it is to some branches of engineering, but a barrier has always separated the two professions. This dichotomy has been detrimental to both groups—and to the beneficiaries of their services.

President Julius W. Adams, in his 1873 inaugural address, described the architect as the perpetuator of art and design that characterized the monumental works of the ancients, works built mainly for the rich. He considered civil engineering to be a modern science that "has grown out of the wants induced by modern ... culture, and the more luxurious ... comforts which, no longer confined to the governing classes, as the conditions of the masses ... bettered have become necessities of life."

President Adams described the relationship between the civil engineer and the architect: "The science of construction in all its ramifications, however essential it may be to the Architect for embodying his ideas, yet, is entirely within the province of the Civil Engineer; whilst the decorative branch of art merely is all that properly belongs to the architect as his specialty."

Alonzo J. Hammond (President, 1933) thought that,

> in a comprehensive sense, Civil Engineering includes Architecture as a Mechanical Art, in distinction to Architecture as a Fine Art. ... Architects in good standing have always been eligible for membership; and especially at the present time, an architect competent to design and execute the complicated details of one of the large buildings now becoming so common in all large cities, is surely the peer of other engineers.

Through the years, the architectural profession has tried to ensure its stature by legislative edict, to ASCE's displeasure. In March 1876, the Society's Board took note of a "Bill to establish a Bureau of Architecture" in the Treasury Department. The legislation, drafted by AIA, was introduced in the House of Representatives. The outcome is not known.

The issue of professional licensing arose in architecture several years before it appeared in engineering. Both AIA and ASCE lost no time in providing guidelines for registration laws, even though both organizations were originally opposed to the idea. The architectural registration laws contained the broadest possible definition of the profession, resulting in barriers to some areas of civil engineering practice, particularly in the design of buildings. Architectural registration boards have vigorously defended the definition, and they have charged many engineers with encroachment into architectural practice although they were performing services for which they were fully qualified by education and experience.

Only occasionally has this reaction been reversed. In 1920, when the Allegheny County Commissioners proposed to employ architects to design and supervise the construction of several important bridges in Pittsburgh, the Association of Members in that city brought the matter to ASCE. The Society protested such action as being "detrimental to the public interest to subordinate safety and economy, adequacy for future traffic, and cost of these structures to their appearance, although ... the ... aesthetic features of bridges may properly be entrusted to those especially skilled in architecture." The American Institute of Consulting Engineers adopted a similar resolution. The joint Committee of Engineering Council and AIA disagreed, saying that "whether the engineer is chief and the architect associate or vice versa is an administrative detail of relative unimportance." ASCE reminded the Engineering Council that it had no authority to represent the Society in a matter solely within the province of the civil engineer, and it widely publicized its resolutions. The issue remained moot until 1923, when ASCE reiterated its position to the County Commissioners. This time, they paid heed.

While the Allegheny County controversy was going on, the joint Committee of Engineering Council and AIA was adopting a resolution addressed to management of licensing problems in the two professions. At the same time, AIA was urged to become a full member of the Engineering Council. Apparently neither action was implemented.

Only rarely have architects been charged with transgression into engineering practice under the licensing laws, possibly because engineers are occasionally engaged by architects.

The definition of the practice of architecture in registration laws was a matter of contention between the professions in the late 1920s. A 1931 joint Committee on Registration of Architects and Engineers "had a marked effect in clarifying the situation with regard to proposed legislation affecting the interest of engineers." However, the effect was temporary.

The effort to use registration law to stake out jurisdictional claims reached a peak in 1971, when the National Council of Architectural Registration Boards (NCARB) promulgated legislative guidelines defining architecture as covering "the design and construction of a structure or a group of structures which have as their principal purpose human habitation or use, and the utilization of space within and surrounding such structures." This definition would have subordinated civil engineering and all other design professions to architecture!

ASCE and other engineering societies protested loudly and alerted their State and Local Sections to proposed amendments to architectural registration laws. A promising result was the creation in 1972 of a joint liaison committee of the NCARB and the National Council of Engineering Examiners (NCEE).

During World War II, the term "Architect–Engineer" came into broad use applied to professional services on large Army projects. In 1945, President J.C. Stevens asked the Army Corps of Engineers to amend this practice, giving the consultant architect or engineer whichever title was appropriate. The Corps acceded readily, but by this time, the term had found a place in civil practice, and it was still in fairly general use in 1974.

On the positive side, an AIA historical document mentions that sometime about 1902, ASCE was one of four bodies to be invited to send nonvoting delegates to the AIA conventions. AIA and ASCE participated in a successful conference of eight engineering, architectural, and construction organizations held in December 1921 to produce a standard form of construction contract.

In 1933, AIA and ASCE created the Engineer–Architect Joint Committee to ameliorate professional differences. An immediate objective was joint study and action on the then-proposed federal Department of Public Works, which both organizations favored.

In 1938, a joint effort of ASCE, AIA, and the Engineering Council sought to improve procedures for employing engineers and architects on public works projects. The result was more liberal policies on the use of private consultants.

Another joint venture in 1941 by ASCE, AIA, ASME, and the American Society of Landscape Architects (ASLA) produced the first attempt to outline the specific areas of professional jurisdiction for civil and mechanical engineering and for architecture and landscape architecture on certain national defense projects. All four organizations adopted the resulting statement, "Division of Responsibility and Work Among the Planning Professions of Architecture, Civil Engineering, Landscape Architecture and Mechanical Engineering on National Defense Housing Projects."

After World War II, the ASCE–AIA joint committee became known briefly as the Joint Committee of the Design Professions, and later as the ASCE–AIA Joint Cooperative Committee. In 1951, the committee helped improve procedures for awarding professional service contracts on defense projects. During the ASCE Centennial year of 1952, AIA featured the "Reunion of Engineering and Architecture" at its convention, a gracious gesture, albeit more fanfare than fact. The cooperative committee sponsored a successful Public Works Conference in Washington in 1956 under the joint aegis of ASCE and AIA.

Around 1958, to bring the point of view of mechanical and electrical engineers into the discussions, the ASCE representation in the joint committee was transferred to the Engineers Joint Council (EJC). This was an unfortunate move; after a few years of lukewarm participation by EJC, the liaison was dropped, leaving ASCE and AIA without a cooperative link. This changed in 1963 when the executive directors of AIA and ASCE decided to form the Interprofessional Commission (later "Council") on Environmental Design (ICED). Along with AIA and ASCE, the original participant organizations included ASLA and the American Institute of Planners (AIP). The Consulting Engineers Council (later the American Consulting Engineers Council), the National Society of Professional Engineers (NSPE), and the American Society of Consulting Planners (ASCP) soon became members.

ICED was intended to provide "a top level mechanism for communication, long-range planning, the definition of major interprofessional problems and objectives and means for their solution and attainment." Council members were the Presidents of each society during their three-year appointments, and each society's Executive Director. Because this ensured a forum at both the policy and executive levels of each society, ICED was immediately productive. The 1941 "Division of Responsibilities" policy was updated to a new guide to "Professional Collaboration in Environmental Design."

The group set up a forum for local mediation of interprofessional controversies, adopted a policy statement on professional criticism of public works projects, and considered problems in ethical standards, registration, and overlapping activities. ASCE managed a successful series of four ICED conferences, covering professional collaboration, environmental design education, social aspects of design, and environmental impact.

Deliberations in ICED concerning the controversial definition of architecture proposed in 1971 resulted in the Interprofessional Council on Registration, comprising the national organizations of registration boards in architecture, landscape architecture, and engineering. The widespread solicitation of political contributions from engineers and architects seeking contracts for public projects greatly concerned ICED in 1974.

ICED showed great potential, but there were setbacks due to unilateral actions by AIA. A controversial film on urban transportation, a rule denying recognition of professional experience acquired by an architect in an engineering firm, and advertisements implying architectural expertise in traditional areas of engineering had bitter reaction from ICED. In 1973, for example, an AIA advertising campaign invited firms to seek architects to solve pollution, traffic, noise, and environmental problems.

Mutual trust and empathy between civil engineers and architects are vital to both professions and their clients. There is a long way to go.

Intersociety Relationships

Interaction among American engineering societies has been cordial over the years, but not necessarily effective. As the number of societies increased to more than 150 national engineering and related organizations in 1972, unification and organization of the profession were more difficult than was the case in medicine and law.

In 1876, ASCE and AIME were the only national societies, and they joined in two ventures: a program for the Philadelphia Centennial Exposition and a joint Committee of ASCE and AIME on Technical Education. In 1885, ASCE, AIME, ASME, and AIEE formed a Committee on a Joint Library, which was unsuccessful in a three-year stint. There was considerable intersociety cooperation for the international expositions in Paris (1889), Chicago (1893), Saint Louis (1904), and San Francisco (1915).

Late in 1889, an intriguing idea was put before the Board of Direction by Henry R. Towne. He proposed an "Institute of Engineers," a supersociety composed of eminent engineers selected by the four Founder Societies. This body would elect from its membership a "Senate" to speak for the entire profession. After deliberation, the Board determined that the plan would "be inexpedient."

For several years after the organization of AIEE in 1884, that group met in ASCE's headquarters, and in 1890, AIEE expressed its gratitude with a gift of fireplace fixtures.

The intersociety John Fritz Medal Board of Award was created in 1902 and was still functioning in 1974.

In 1907, the Society for the Promotion of Engineering Education (which later became the American Society for Engineering Education, ASEE) brought the "Big Four" together in a joint Committee on Engineering Education, which the Carnegie Foundation later funded. Its final report was made in 1919. Also in 1907, the four societies each selected one member to form a library committee. The group was instrumental in establishing the Engineering Societies Library in 1915. In the meantime, AIME, ASME, and AIEE set up the United Engineering Society in 1904 as a corporation to administer a gift from Andrew Carnegie (see Chapter 3) "to be used for advancement of the engineering arts and sciences, and all their branches, and for the maintenance of a free public engineering library." In 1916, ASCE became the fourth Founder Society and joined the other three in the Engineering Societies Building. (The American Institute of Chemical Engineers, AIChE, was admitted as the fifth Founder Society in 1958, when the United Engineering Center project was undertaken.)

The proposal to form an "Academy of Engineers" in 1918 elicited this response from ASCE: "Such an organization ... if there is a field for its activities, should be organized under the initiative of the National Engineering Societies, and ... that the membership of an Academy of Engineers should not be made up by self-appointment."

The Board of Direction reacted again in 1924 to the news that an American Society of Engineers was being formed in the Chicago area. They requested the Society's legal counsel to protest against the similarity of the proposed name and badge to those of ASCE. The matter appears to have been closed with this action.

Engineering Society Federations

Up to about 1912, intersociety collaboration was strictly ad hoc. It appears that the earliest continuing cooperative arrangement was a General Conference Committee of the National Engineering Societies, which reported in 1913–14 on several professional matters. In April 1917, this rather informal general joint committee created in the United Engineering Society a new department called the Engineering Council, constituted initially of five representatives from each Founder Society, "in order to provide for convenient cooperation between the Founder Societies, for the proper consideration of questions of general interest to engineers and to the public and to provide the means for united action upon questions of common concern to engineers."

This was the first formal federation of engineering societies in the United States. The Committee provided for admitting other approved engineering or technical organizations, and the American Society for Testing Materials and the American Railway Engineering Association joined the group.

During World War I, the Engineering Council of National Technical Societies had many opportunities to serve the public and the profession, and it did so with distinction. The Engineering Council office opened in Washington, D.C., in 1918, and its activities ranged from legislation to cooperation with many federal agencies. They produced a roster, "Civil Engineers for War Service," as part of a comprehensive engineering labor survey. In December 1918, the Engineering Council set up the Engineering Societies Service Bureau, an employment service for members, with the Founder Society secretaries as the managing committee.

Then came the first of many obstacles to organizing the engineering profession during the next half-century. Three Founders, AIME, ASME, and AIEE, decided that an autonomous structure was desirable, and in 1920 they formed the Federated American Engineering Societies (FAES), with the American Engineering Council as its executive arm. Despite ASCE's preference for continuing the Engineering Council, it was dissolved in December 1920 when the proponents of the new federation withdrew their support.

An ASCE questionnaire elicited an expression of almost 60% against affiliation with FAES. The Society's refusal to become a Charter Member of FAES appeared to stem from legal questions and a reluctance to assume the $11,000 annual financial commitment. However, the Board was in a dilemma. On the one hand, the membership endorsed the recommendation of the 1919 Committee on Development "to actively cooperate with other engineering and allied technical organizations in promoting the welfare of the Engineering Professions." On the other, there was no acceptable medium available for such cooperation. A Committee on External Relations appointed to resolve the problem concluded that intersociety cooperation "shall not involve the surrender of the name and standing of the Society into outside hands, or in any way permit the use of the Society's name in behalf of any cause of which it does not approve."

ASCE sought unsuccessfully, in 1922, to resolve the impasse by asking the Engineering Foundation to study "the existing confusion and duplication of activities in Civil Engineering organizations." The foundation tabled the proposal when it appeared that it would not be supported outside the ASCE delegation.

After an ASCE membership referendum in 1923 affirmed the decision to forgo affiliation with FAES, the Society enlisted the other three Founders in a joint Committee on Cooperation in Public Matters. The Presidents and Secretaries of the Four Founders established a continuing forum for interaction. The joint Conference Committee was highly effective, and it continued for seven years, enabling ASCE to support such American Engineering Council causes as it chose. The Society also supported the Engineering Societies Service Bureau when it was under the direction of FAES.

In 1924, FAES took the name of its executive body, the American Engineering Council. Three years later, ASCE inquired about the desirability of certain amendments to the constitution, bylaws, and rules of the council. This led to another referendum in the Society in 1923, this time overwhelmingly favorable to affiliation.

At this stage, the American Engineering Council comprised six national, five state, and 15 local societies. Its interests were strongly oriented to those of ASCE in such areas as reforestation, topographic surveys, national hydraulic laboratory, public health engineering, federal public works administration, water resources, highway safety, and public policy on power. Many other research studies and programs were directed toward industrial problems and appraisals.

The American Engineering Council was active through the early years of the Depression, but in 1934 the lack of contributions from member organizations led to staff and program reductions. By July 1940, fiscal problems and general apathy were such that ASCE gave notice of its withdrawal from the council. The council's leadership urged reconsideration, but to no avail; it was dissolved on January 1, 1941.

A few months later, ASCE invited the other Founders to join in setting up a Washington office. This did not develop, but the Founders and AIChE did reestablish a joint Conference Committee, with the president and secretary of each society as delegates. The committee provided intersociety communication and liaison from 1923 to 1930, when it was dissolved during the heyday of the American Engineering Council.

Once again, the joint Conference Committee, unhampered by administrative and hierarchical ballast, accomplished much. The committee oversaw an in-depth study of the engineering profession and the economic status of engineers, and, in 1945 and 1946, produced reports on the industrial disarmament of Germany and Japan after World War II.

Engineers Joint Council

The joint Conference Committee changed its name to the Engineers Joint Council in 1945, and it added international relations to its undertakings. The 1946 survey report, "The Engineering Profession in Transition," was the most comprehensive statistical study ever made of the engineering profession to that time. The Engineering Manpower Commission was initiated as an EJC function in 1950. By 1952, the Exploratory Group recommended that EJC be reorganized, with delegates to be active members of the governing body of the representative societies. The move was made, EJC membership was expanded to 10 organizations, and programs were successfully developed and pursued.

The unity movement continued to elicit many proposals and charts for restructuring the profession. Several EJC societies were concerned about their tax-exempt status in view of the noneducational professional nature of most EJC programs. Some leaders in EJC societies were also active in the NSPE (organized in 1934) and sought to establish that society as the professional representative for all of engineering. In 1958, ASCE issued a policy statement:

> In ASCE, unity means cooperative effort in the common interest. The Society welcomes the opportunity to work with other organizations in the attack on and solutions of common problems. At the same time, ASCE continues to apply itself to the same activities and problems in advancing the special interests of the civil engineer. ...
>
> Envisaging the great contribution to the profession that could result from the participation of [NSPE] in both Engineers Joint Council and Engineers Council for Professional Development, [ASCE] would view with high favor the simultaneous affiliation of NSPE with both EJC and ECPD.

NSPE would not accept membership in EJC, however, so this plan for "unity" was also unsuccessful.

EJC achieved its greatest visibility in 1960 with a new quarterly newsletter, *Engineer*, as the result of an ASCE proposal. Federation productivity was at a high level at this time, but the restiveness of some constituent societies began to tell. In 1964, the National Academy of Engineering (NAE) began operating as a counterpart of the National Academy of Sciences. The objectives of NAE impinged somewhat on those of EJC, and three years later EJC underwent another and disastrous reorganization.

This time EJC was placed under the direction of a Board made up of engineers in responsible positions in industry but with almost no past or present relationship to the management of the constituent societies. IEEE, AIChE, and the American Society of Heating, Refrigeration, and Air-Conditioning Engineers all withdrew from EJC membership in 1968, though some readjustments in organizational structure were hastily made.

In 1970, ASCE—then the largest of all EJC societies—reviewed its own membership status. The Society decided to continue as a member of EJC and to help in rebuilding it as an effective forum and cooperative federation for the engineering profession.

United Engineering Trustees

The United Engineering Society, formed in 1904 by AIME, ASME, and AIEE and joined later by ASCE and AIChE, proved to be the most consistently viable of the older engineering federations. This is no doubt due to its sole purpose—to administer as trustee the funds contributed for the general benefit of the engineering profession. Its primary function had been to build and operate the Engineering Societies Building from 1906 to 1961 and the United Engineering Center in New York City after 1961.

In 1914, the United Engineering Society created a subsidiary, the Engineering Foundation, to administer substantial gifts made by Ambrose Swasey, a Past-President of ASME, for research and advancement of the profession. The Engineering Societies Library also became a part of the United Engineering Society in 1915, although ASCE did not surrender its own library until a year later. In 1931, the United Engineering Society changed its name to United Engineering Trustees, Incorporated.

The Engineering Societies Personnel Service

A joint personnel service was an important intersociety activity from 1918 until 1965. After the end of World War I, the Engineering Societies Service Bureau helped engineers who had served in the armed forces find employment. The Engineering Council reluctantly directed the bureau, though it was being funded by the Four Founder Societies and was managed by the Founder Society secretaries.

Engineers used the bureau without charge. In 1919, there were 1,256 placements of 5,377 registrants, and in 1920 placements totaled 1,606 from 2,256 registrants, at an average cost of $7.93. Local offices were opened in San Francisco and New York. Direction of the bureau was assumed by the FAES in 1921, but 18 months later reverted to the Founder Societies. ASCE continued to pay its share of the subsidy even though it was not a member of FAES.

In 1923, the bureau was established as a cooperative undertaking of the Founders with management still under the committee of secretaries, and it was put on a fee basis with half the operating cost still subsidized. A third local office was opened in Chicago.

The bureau continued into the Depression decade of the 1930s, performing modestly but well. In 1932, there were 3,520 registrations and 573 placements. Its usefulness increased as the Depression waned, and in 1941, it became the Engineering Societies Personnel Service, Incorporated (ESPS). Local offices were opened temporarily in Detroit and Boston. Soon, income from fees exceeded expenses, and ESPS accumulated a substantial reserve. After World War II, the employment situation for engineers changed for the better. ESPS's reserve funds were diminished by operating deficits, and the operation was terminated in 1965.

Since its organization in 1930, the Engineers' Council for Professional Development (ECPD; see Chapter 4) has been one of the most successful of all the engineering federations, probably because its objectives are well defined and are shared by all segments of the profession. ECPD is the accreditation medium for engineering education, and ASCE has wholeheartedly supported the group since its formation.

The Continuing Quest for Unity

The idea of a consolidated engineering profession changed in the latter 1950s to a concept of a super-organization formed by merging of federations. There had been an earlier proposal to merge EJC and ECPD into an "American Engineering Association, Incorporated," within which the functions of the two federations would be served by separate operating divisions. NSPE would have participated, and the plan was approved by ASCE in February 1959 but never consummated.

The Board of Engineering Cooperation (BEC) evolved from this idea in 1960. BEC was an informal assembly of the Presidents, Past-Presidents, and Presidents-Elect of EJC, ECPD, and NSPE. Occasional meetings were held to review the goals and activities of the three organizations, to enhance cooperation, and to minimize duplication. In 1970, ASEE became the fourth partner in BEC.

Early in the 1960s, BEC decided to invite the Presidents of the EJC constituent societies to join them in one meeting each year, and it later invited the entire executive committee. The Joint Societies Forum, as it became known, specifically excluded any staff representation.

BEC and its yearly Joint Societies Forum hoped to be the launching pad for an umbrella organization for the entire profession. In 1968, the oft-proposed concept of an individual membership organization for all engineers was discussed for a few months. The certification concept, which had been promoted by ECPD in 1934, was resurrected in 1969 in a plan for "accrediting" engineers who did not choose or who might not be able to qualify for registration. The standards and evaluation of qualifications would have been made by a new "Institute for the Accreditation of Engineers."

The Joint Societies Forum referred the proposal to EJC in 1971, and three years later it was still under study by an EJC task force. ASCE was uncommitted on the matter in 1974. In August 1973, the Joint Societies Forum recommended that an Ad Hoc Committee on Unification in Engineering be set up to devise a new organizational structure that would realize "a unified approach to many problems confronting the profession and the nation … beneficial to individual engineers and the public." BEC accepted the challenge, and in 1974 the movement to capture and contain the elusive specter of engineering unity was again in full force, past failures notwithstanding.

ASCE has a long list of external associations. Communication among the Founders grew after ASCE moved into the Engineering Societies Building in 1917. Joint committees were set up, such as the joint ASCE–AIME Committee on Technical Education (1876) and the ASCE–AIChE Committee on Water Pollution (1932–38). The spontaneous exchanges and collaboration, however, were far more productive. The Founders were very good neighbors, even though they did not always agree.

Outside the Founder group, ASCE had contact of some kind with all national organizations engaged with specialty fields of civil engineering, and a host of others. Descriptions and history of 10 of these follow.

Highway Research Board

In 1920, the Engineering Division of the National Research Council created the Highway Research Board (HRB) to carry out a national program of highway research, and it invited ASCE to participate along with 11 other groups. The Society endorsed the proposal, pledging its assistance in raising funds and in publicity. In 1974, HRB changed its name to the Transportation Research Board. ASCE'S cooperation has continued through its Highway Division.

Chi Epsilon Fraternity

The scholastic honor society in civil engineering, Chi Epsilon, was formed in 1922 by students at the University of Illinois. One of the founders, Harold T. Larsen, devoted most of his career to ASCE's publications staff and retired as Manager of Technical Publications in 1963. When ASCE occupied its new offices in the United Engineering Center in 1961, Chi Epsilon conducted a special fund-raising drive and contributed $15,000 to the Society to furnish its new conference room, the Chi Epsilon Room.

Associated General Contractors of America

ASCE collaboration with the Associated General Contractors of America (AGC) began more than a half-century ago in the joint Conference of Engineers, Architects, and Constructors, which worked on developing a standard form of contract for engineering construction. The venture was conceived by the ASCE Committee on Highway Engineering in 1921, and the first edition of the "Standard Contract for Engineering Construction" was published four years later. By that time, in addition to ASCE and AGC, the joint conference included the American Association of State Highway Officials (AASHO), the American Engineering Council, AIA, the American Railway Engineering Association (AREA), the American Water Works Association (AWWA), and the Western Society of Engineers.

In 1923, AGC and ASCE discussed the possibility of a buyers' strike in protest against the rising costs of construction at that time. AGC produced a report, "The Seasonal Demand of Construction for Labor, Materials and Transportation," and the Society asked its Sections to use this document at local meetings on "Waste in the Building Industry" and to express their views on ASCE's forming a Construction Division. The new Division began in 1925. Since 1948, the ASCE–AGC joint conference has provided liaison between the two bodies.

Water Pollution Control Federation

In 1927, ASCE's Sanitary Engineering Division directed a task committee to explore ways to disseminate knowledge on sewage and industrial waste treatment and on pollution control. This committee called a general meeting in

Chicago on June of that year, bringing together representatives of AWWA, the American Public Health Association (APHA), the Conference of State Sanitary Engineers, and the American Society for Municipal Improvements. Charles A. Emerson Jr. became the chair of a Committee of One Hundred. These efforts led in January 1928 to the Federation of Sewage Works Associations. In 1960, the Federation became the Water Pollution Control Federation (WPCF).

Construction League of the United States

The intent of the Construction League of the United States was to mobilize the construction industry during the Depression. The League, formed in 1932, included ASCE as a charter member along with AIA, AGC, the American Road Builders Association (ARBA), and the American Institute of Steel Construction (AISC). Some 15 other technical and trade associations were also members.

The Construction League drafted the Construction Industry Code of Fair Competition, procedures for coordinating the industry. President Franklin Roosevelt approved Chapter 1 early in 1934, which included guidelines on bidding said to be "a lasting norm of fair competition." The National Recovery Administration (NRA) administered the code. Another chapter of the Fair Practice Code, drafted by the League's Engineering Division, was the so-called Engineering Code, which was published in the June 1934 *Civil Engineering* but apparently never approved by NRA. The league also provided input from participants relating to public policy and legislation.

National Society of Professional Engineers

Several prominent ASCE members helped organize NSPE in 1934, but the movement was in no way sanctioned officially by ASCE. On the contrary, the ASCE Board of Direction directed the Secretary in January 1935 to issue a statement making it clear "that the Society has no official representative in the newly formed National Society of Professional Engineers."

Among the leaders in ASCE at this time were some who did not see the need for a new organization "devoted exclusively to the professional interests of all engineers." The other Founder Societies were quite willing to accept NSPE, and the newcomer prospered in the post-Depression demand for stronger economic welfare programs for engineers.

NSPE membership was limited to registered engineers for 36 years, and as far more civil engineers were registered than other engineers, there was substantial duplication of membership. The relationship between the two was characterized by a restrained aloofness, with some competition in professional programs.

By 1955, the barriers between the two societies diminished, and cooperation began in the mid-1940s when NSPE joined with ASCE and other bodies in supporting the professional employee provisions of the Taft-Hartley Act. The relationship continued to improve through the 1960s, with day-to-day staff interaction and increasing joint representations in such bodies as ICED, the Committee on Federal Procurement of Architect–Engineer Services (COFPAES), ECPD, NCEE, and other task committees. In 1974, the two societies were cordial and collaborative in every respect.

The 1934 activities directed at that time toward organization of NSPE and other new engineering bodies had considerable impact on the course of ASCE. The Board set up a task force to analyze the objectives of these movements, and this might explain the Society increased efforts at improving engineering salaries in the late 1930s through the 1940s.

American Academy of Environmental Engineers

The American Academy of Environmental Engineers is the first and the only organization (in 1974) with a roster of engineers certified in specialty field.

In the early 1950s, ASCE's Committee on Advancement of Sanitary Engineering noted that sanitary engineers engaged in the public health field needed to define their level of competency to compete with medical public health personnel for ranking administrative positions. Because sanitary engineering is interdisciplinary, the ASCE committee expanded in 1952, becoming the joint Committee for the Advancement of Sanitary Engineering, and included APHA, ASEE, AWWA, and the Federation of Sewage and Industrial Wastes Associations (later WPCF).

The American Sanitary Engineering Intersociety Board, Incorporated, emerged from this group, founded in 1955 patterned after that of the American College of Surgeons. This Board set up criteria and procedures for certifying registered engineers on the basis of their education, experience, and an examination, for proficiency in sanitary engineering. Successful candidates became diplomates of the American Academy of Sanitary Engineers.

ASCE supported the Intersociety Board by a loan of $7,500 to cover start-up costs and provided complimentary office facilities from 1955 to 1958 at the headquarters of the Society. In the mid-1960s, the names of the board and academy were changed to the American Environmental Engineering Intersociety Board, Incorporated, and the American Academy of Environmental Engineers. In 1973, the board was merged into the academy. This made the academy essentially autonomous, although its 18 trustees included representatives of ASCE, AIChE, APHA, ASEE, AWWA, WPCF, the American Pollution Control Association, and the American Public Works Association (APWA).

American Consulting Engineers Council

The American Consulting Engineers Council (ACEC) is the result of a 1973 merger of the American Institute of Consulting Engineers (AICE) and the Consulting Engineers Council of the U.S.A. (CEC). AICE, formed in 1910, was a small but prestigious group of prominent consulting engineers. Its membership reached about 450, most of whom were also members of ASCE. Both groups were interested in ethical matters and in public issues impinging on private practice, and the relationship was always cordial and highly professional.

CEC was founded in 1959 to fill the need for an association to further the promotional and business techniques of the consulting engineer, as is done by AIA for the architect. Its membership was limited to firms, either partnerships or corporations.

When CEC suggested to the Society in 1960 that a joint liaison committee be set up, the Board recognized that CEC filled a need that could not properly be served by ASCE. A three-way liaison, comprising CEC, AICE, and ASCE—the "Tripartite Committee"—continued until 1965, when its purpose was fulfilled within the Coordinating Committee on Relations of Engineers in Private Practice and Government (later COFPAES). In the meantime, CEC grew to a membership of 2,500 firms.

In the late 1960s, there was a movement to combine AICE, CEC, and the Professional Engineers in Private Practice (PEPP) Division of NSPE. PEPP abstained, and the 1973 merger action brought AICE and CEC together as ACEC.

Committee on Federal Procurement of Architect–Engineer Services

Some federal construction agencies insisted on competitive bidding for professional services, and this led in 1964 to the formation of the Coordinating Committee on Relations of Engineers in Private Practice with Government. ASCE was a participant, along with AICE, CEC, NSPE, and ARBA. In 1968, AIA joined, and the committee became COFPAES. This joint venture assisted a Congressional task force appointed in 1967 to study architect–engineer contract negotiation procedures, and it also afforded a communication forum during the antitrust actions by the Department of Justice against ASCE and AIA in 1971–72. Its major accomplishment was the original drafting and support of the 1972 Brooks Bill, which required all professional services for construction projects involving federal funds to be contracted by professional negotiation.

COFPAES faced an obstacle when 1973 disclosures of political contributions by engineers in Maryland resulted in the resignation of the Vice President of the United States. Some public officials and newspapers took the view that selecting professional consultants by competitive bidding would preclude such corruption.

American Society of Certified Engineering Technicians

In 1967, the Board of Direction resolved to recognize the civil engineering technician, noting that the engineer and technician should preserve their independent identities, high standards of competency, and organizational autonomy, while cooperating and providing maximum service to members. The ASCE–American Society of Certified Engineering Technicians (ASCET) joint Committee was formed in 1968. The "Civil Engineering Technician Service

Roster," was an innovation. On payment of a modest fee, any ASCET member could receive ASCE publications and meeting notices. The plan emphasized ASCE–ASCET interaction at the local level, and it was in force in 1974.

ASCE has been part of standing "Joint Cooperative" committees over the years, including those with AGC, ASCET, and with AIA from 1933 to 1968. More recent joint committees have been with AIP during the period 1959–63, AREA in 1966, and APWA in 1968.

There have been close working relationships with other engineering and related bodies without formal organizational nomenclature. Among these are ARBA, AWWA, American Concrete Institute, Prestressed Concrete Institute, AISC, American Iron and Steel Institute, Western Society of Engineers, AASHO, American Arbitration Association, American Geophysical Union, American Institute of Aeronautics and Astronautics, American Nuclear Society, Institute of Traffic Engineers, and many, many others. ASCE's "cooperation in the common interest" with other groups also took place through Local Sections and the Technical Divisions.

International Engineering Relationships

ASCE has had a basic role in the development of the engineering profession in the nation, but the Society has been equally influential in furthering engineering relationships throughout the world.

The Corresponding Member grade authorized in the original Constitution gave ASCE an international dimension from its beginning. Such members were nonresidents of the United States, knowledgeable in an engineering specialty. They paid no dues but had to correspond with the Society at least once a year.

The first international interaction was in 1874, when the Association of Engineers and Architects of Austria proposed a publications exchange and invited ASCE members to visit its Vienna headquarters. The Society accepted, and it extended courtesy to foreign engineers attending the Philadelphia Centennial Exposition in 1876. Cards of introduction and a "List of Members" were given to accredited members of foreign societies.

After the International Engineering Congresses in Paris in 1889, Chicago in 1893, Paris in 1900, St. Louis in 1904, and San Francisco in 1915, there were a series of exchange visits. Fifty-two ASCE members and 33 wives visited France, England, and Germany for the 1889 celebration of the Centennial of the French Republic in Paris.

Two events in England were highlights: an Institution of Civil Engineers dinner in the London Guildhall and "the special permission given by Her Most Gracious Majesty, the Queen, to visit and inspect her Royal Palaces and Domains at Windsor and in the Metropolis." The Institution of Civil Engineers invited the Society to hold a full-scale convention in the London headquarters of the Institution after the Paris Exposition of 1900. After being hosted by the Societe des Ingenieurs Civiles de France in Paris, 68 ASCE members moved on to London for the 32nd Annual Convention of the Society. Her Majesty, the Queen, received the group at Windsor Castle, as did the Earl and Duchess of Warwick at Warwick Castle. Again, the Institution of Civil Engineers (ICE) held a reception at the City of London Guildhall.

ASCE went all out to reciprocate for foreign guests attending the International Engineering Congress arranged and financed solely by ASCE in connection with the 1904 Louisiana Purchase Exposition. The Society gave each visitor a letter of introduction, and it set up host committees in 16 major cities to accommodate the guests. A delegation of 104 members and guests from ICE accepted the invitation of the Society to take part in a four-day program of technical visits and excursions in and near New York City, and a dinner at Delmonico's Restaurant.

In 1905, the Board of Direction authorized a communication "to all engineering societies, both in this country and abroad, stating that this Society would be glad to welcome any of their members to the use of this House and Library, and to any of its regular meetings." Sixteen foreign societies promptly accepted and reciprocated these courtesies. The *Annual Yearbook* listed them until 1949, by which time 30 foreign societies were included.

A 1950 Bylaw provided for a reciprocal membership agreement, for the Secretary or equivalent officer, with the Founder Societies and a list of 20 engineering societies in Europe and Latin America. The idea was sound, but the purpose was better served through the personal contacts made possible later by international organizations.

An open policy on the establishment of ASCE Sections in foreign countries prevailed until 1958. Sections at that time were in Panama (1931), Venezuela (1947), Brazil (1948), Mexico (1949), and Colombia (1957).

After a policy review in 1958, ASCE decided that it should encourage its members abroad to become a part of the engineering community of the country of their residence and that the Society should not compete with national engineering societies abroad. This led to two important developments: a decision to suspend any further authorization of ASCE Sections outside the United States, and a requirement that a candidate for membership from a foreign country must hold the equivalent grade of membership in a recognized national engineering society in his own country.

An informal "Overseas Unit" plan was introduced on a trial basis in Australia and Pakistan only for the identification and communication of ASCE members in those countries. This was not too successful. The Sections in Brazil and Venezuela were liquidated in the early 1960s.

The Institution of Civil Engineers

ICE was founded in 1818, receiving its royal charter 10 years later. It had some 700 members when ASCE appeared on the scene, and it was a model for its American counterpart.

ICE made an early gesture of friendship to ASCE, contributing the first 20 volumes of the ICE *Proceedings* to the Society's library in 1882. The exchanges of visiting delegations during several international congresses between 1889 and 1915 established rapport that has prevailed ever since.

In the early 1900s, the Institution held periodic General Engineering Conferences. The first of these held after World War I in 1921 and a delegation from the four American Founder Societies attended. At this time the John Fritz Medal was conferred upon Sir Robert A. Hadfield in London and upon Charles P.E. Schneider in Paris.

The New York World's Fair in 1939 was to have been the setting for a British–American Engineering Congress, as a joint venture of ASCE, ICE, and the Engineering Institute of Canada. The event had to be canceled after the program was arranged due to the precarious situation in Europe in September 1939. After 1949, ASCE and ICE interacted at international federations such as Conference of Engineering Societies of Western Europe and the U.S.A., known as EUSEC, and the World Federation of Engineering Organizations (WFEO). One result was the 1969 Conference on World Airports in London, in which the ASCE Aero-Space Transport Division took part. Another was a plan to hold biennial joint conferences on topics of mutual concern. The first ICE–ASCE joint conference was held in 1970 in Bermuda, on the theme "The Engineer in the Community," and the second, at Disney World, Lake Buena Vista, Florida, in 1972, on the theme "Public Works and Society." Both conferences produced published proceedings of high reference value.

Société des Ingénieurs Civiles de France

Société des Ingénieurs Civiles de France (SICF) hosted ASCE during the Paris Expositions in 1889 and 1900, and the French Congress General du Genie Civil in 1919. The latter event led to a permanent Franco–American Engineering Committee. The ASCE Board of Direction endorsed the new committee and named members to serve with representatives of the other Founder Societies, which apparently were less enthusiastic. The *Annual Yearbook* of ASCE listed representatives on the committee until 1929, but there is no record of activity.

There was an American Section of the SICF in New York City some time prior to 1923. SCIF's 75th anniversary in 1923 was celebrated by the American Section in a joint meeting with the Founder Societies. SCIF was still active in 1974.

The Engineering Institute of Canada

Only about a dozen Canadian civil engineers were on the ASCE roster in 1881 when the Society held its thirteenth Annual Convention in Montreal. The Canadian Society of Civil Engineers (CSCE) was founded in 1887. Although reciprocal courtesies to visiting members with ASCE began in

1908, there were no joint activities until 1917. ASCE held Annual Conventions in Quebec in 1897 and in Ottawa in 1913, and CSCE appeared to have helped with staffing local arrangements committees.

The onset of World War I gave rise to a 1917 resolution by the ASCE Board of Direction stating that "the two Societies should cooperate for mutual advancement to the greatest extent possible" and further recording "approval of a plan of holding joint meetings of the two Societies at such times and places as ... convenient." This gesture was acknowledged by the Canadian Society, but the effort was apparently superseded by the exigencies of World War I.

In 1918, CSCE became the Engineering Institute of Canada (EIC), under the provisions of a new federal charter. When ASCE held a convention in Montreal in 1925, the Montreal Branch of the Institute extended hospitality. The first full-scale joint convention was held in Victoria in 1934, followed by successful meetings in Niagara Falls in 1942, Toronto in 1950, and Montreal in 1974. EIC also participated in the planned 1939 British–American Engineering Congress, which was cancelled because of events leading to World War II.

In 1968, a Conference on Great Lakes Water Resources was staged jointly by EIC and ASCE in Toronto, focusing on international problems.

For more than 50 years EIC and ASCE have exchanged representation by their top officers at their annual meetings. ASCE did not establish Local Sections or Student Chapters in Canada, in accordance with its international relations policy adopted in 1966. However, ASCE joined with EIC and the British ICE to subsidize modestly the Toronto Area joint Civil Engineering Group, which served the Toronto members of the participating societies.

In 1969, EIC had an extensive reorganization, which created within the Institute "Societies" in major branches of engineering. These were formed for mechanical engineering in 1970, civil engineering and geotechnology in 1972, and electrical engineering in 1973. Almost 1,300 ASCE members were residents of Canada at that time.

Latin-American Relationships

The Founder Societies in 1916 appointed a Pan-American joint Committee "to try and form some closer relationship between engineers of North and South America." Once again, the threat of World War I intervened. Overtures to various societies in South America were not encouraging, and the committee disbanded.

Six years later, Fred Lavis, secretary of the 1916 joint committee, urged that a delegation of North American engineers make an organizational effort at the 1922 Brazilian Centennial. ASCE Secretary Dunlap conferred with Lavis and representatives of the other Founder Societies, and their efforts resulted in significant North American participation in the Rio de Janeiro Congress.

In 1942, the Founders again formed a Committee on Inter-American Engineering Cooperation. A year later, ASCE pledged its cooperation with the engineering societies of South America in biennial or triennial hemispherical engineering congresses, and the result was a 1949 convention in São Paulo.

Finally, on April 19–22, 1951, the Union Panamericana des Asociaciones des Ingenieros (UPADI) held its first convention in Havana. The constituent societies of EJC with EIC represented North America, and 12 national societies represented Latin America. The constitutional aims of UPADI were, among others, to "encourage, promote, expand and guide the work of the engineers of the Western Hemisphere; to encourage the holding of periodical Pan-American engineering conventions; and to promote individual and collective visits."

UPADI held 12 biennial conventions from 1951 to 1974 with strong participation by ASCE. From 1965 to 1972, the Executive Director was a member of the UPADI Board of Directors, and the Society supported every proposal toward engineering comity in the Western Hemisphere through the medium of UPADI.

Official representation in UPADI was limited to only two North American entities (EJC and EIC), as compared with as many as 23 Latin American organizations. The programs and meetings reflected this imbalance, but the majority of the operating budget was borne by the North American bodies. ASCE in 1969 urged greater technological and professional input from North America.

European Affiliations

After World War II, Europeans perceived the need for communication and cooperation among the national engineering societies. To this end, the three chartered engineering institutions in Great Britain, ICE, IME, and IEE, issued an invitation to 14 societies to convene in London in October 1948. This assembly led to the Conference of Engineering Societies of Western Europe and the U.S.A., known as EUSEC—an acronym that soon came to identify a dynamic agency for international rapprochement in engineering.

ASCE, ASME, AIEE, AIME, and AIChE were charter members of EUSEC, together with the three British institutions and national societies in Austria, Belgium, Denmark, Finland (two societies), France, West Germany, Italy, the Netherlands, Norway, Sweden, and Switzerland. Greece, Ireland, Portugal, and Spain affiliated later. The EUSEC bylaws specified that the president and secretary of each society would serve as official delegates, which brought together the policy leaders and the continuing executive officers.

EUSEC's accomplishments in its 22 years covered the gamut of engineering education, standards of professional practice, information documentation and exchange, manpower studies, training of technicians, exchange of membership services, international registration practices, and so on. In addition, participation in EUSEC affairs created personal acquaintanceship and day-to-day cooperation among the executive officers of the societies to an extent never known before.

One professional contribution of EUSEC stood out above all others. This was the 1961 *Report on Education and Training of Professional Engineers*, a comparative study of systems of engineering education in 19 countries, and the most comprehensive study of its type up to 1974. Dr. Thorndike Saville, Hon.M.ASCE, was chairman of the committee and its secretary was the Executive Secretary of ASCE. The latter negotiated a $30,000 Ford Foundation grant to finance the project. Matching grants came from the Organization for Economic Cooperation and Development and the EUSEC societies.

EUSEC held biennial General Assemblies, and the 1958 assembly was in New York. In 1956, the Executive Secretary of ASCE became General Secretary of EUSEC, filling this post until 1963, when he was elected chairman of the federation. In 1968, UNESCO proposed that the governing council of EUSEC be the nucleus of an effort to create a world federation of engineering organizations along its own pattern. It was understood that EUSEC would be disestablished when the viability of a world federation was demonstrated. This took place in 1971.

Worldwide Organizations

The 1921 visit to England and France by members of the American Founder Societies sparked the idea of an infrastructure to bring together engineers throughout the world. The Federated American Engineering Societies put a plan for a world engineering federation before the 1929 World Engineering Congress in Tokyo. The proposal resurfaced after World War II at the World Engineering Conference in Paris in 1946. ASCE, ASME, and AIChE acted in 1947 to set up a U.S. National Committee of the World Engineering Conference, under the auspices of EJC. At this point, the venture subsided, for reasons unknown.

The EUSEC Advisory Committee took up the idea again, with the support of the United Nations Educational, Scientific, and Cultural Organization (UNESCO). In 1966, the first meeting of the Organizing Commission was held in Paris, with Eastern Europe and Latin America represented together with the EUSEC countries. The Executive Secretary of ASCE was designated chair of the commission.

The Organizing Commission met twice to develop a draft constitution. In March 1968, WFEO held its inaugural General Assembly, with 61 nations represented. ASCE's Executive Secretary gave the keynote address and was elected to the WFEO Executive Committee for a four-year tenure.

With its secretariat established in the Institution of Electrical Engineers in London, WFEO assumed the role filled so well by EUSEC for its constituent societies. Successful General Assemblies were held in Paris in 1969, Varna, Bulgaria, in 1971, and New York in 1973. At the 1973 meeting, there was still some uncertainty as to the relationship between WFEO and the regional international bodies such as UPADI and the Commonwealth Engineering Council of the United Kingdom. However, WFEO appeared in 1974 to be soundly founded.

In 1963, ASCE initiated an Intersociety Agreement under which a national society, for a modest fee, would extend certain privileges to visiting engineers from other countries. These privileges included (1) registration in a Guest Mailing List and invitation to all national and local meetings of the host society, (2) an identification card or "Professional Passport" for introduction purposes, and (3) a subscription to the regular periodical magazine of the host society. The plan was reciprocal among the signatories.

In 1974, ASCE had 18 Intersociety Agreements in force with national societies around the world. The principle of this ASCE innovation was endorsed by EUSEC, and it has been adopted by a number of societies abroad.

ASCE has welcomed every opportunity to collaborate with several international technical federations engaged in civil engineering. Some liaisons go back to the early 1900s, such as those with the International Navigation Conference and the International Association for Testing Materials. In 1923, the Society assigned a task force "to formulate a basis of cooperation between International Technical Bodies."

Many U.S. national committees of international associations have begun in ASCE's Technical Divisions. The Society has been actively represented in such international groups as the World Energy (formerly Power) Conference, Commission on Large Dams, International Association for Theoretical and Applied Mechanics, International Association for Bridge and Structural Engineers, International Association for Soil Mechanics and Foundation Engineering, International Association for Water Pollution Research, International Association for Earthquake Engineering, International Association for Shell Structures, and International Commission on Irrigation and Drainage.

A young member of the Society, after a series of technical visits in Europe, reported on the "bridge of cooperation" built by the international activities of ASCE. The Society introduction "was to have remarkable significance; it would have a magical effect across Europe, opening doors and leading to the right people."

Back in 1903, a group of Rochester members reacted strongly to the proposed joint Engineering Building, in part as follows:

> ASCE is recognized by the whole civilized world as a power. We have our library and our property and are self-supporting, and it does not strike [us] favorably that we should lower our standard of qualifications and throw all this into a common fund to be participated in by all societies, however ephemeral or social or scientific, that may from time to time come into existence.

In 1958, the attitude of ASCE was quite different. A Society policy began: "In ASCE, unity means cooperative effort in the common interest. The Society welcomes the opportunity to work with other organizations in the attack on and solution to common problems."

Over the years, the gregarious nature of ASCE in its intersociety relationships transcended its occasional aloofness. The original reluctance of the Society to affiliate with United Engineering Society (1903) and the Federated American Engineering Societies (1920) was more than countered by its subsequent involvement in the successor organizations, and in many others.

Many technical specialty organizations trace their origin to ASCE support, as does the multidisciplinary body, ICED. No other engineering organization in the world surpassed the ASCE contribution to international cooperation between engineering societies. By 1974, the Society had become devoted to "cooperative effort in the common interest" as a way of life.

APPENDIX

Roster of ASCE Presidents, 1853–2002

1853–67	Laurie, James	1896	Clarke, Thomas Curtis
1868	Kirkwood, James Pugh	1897	Harrod, Benjamin Morgan
1869	McAlpine, William Jarvis	1898	Fteley, Alphonse
1870–71	Craven, Alfred Wingate	1899	FitzGerald, Desmond
1872–73	Allen, Horatio	1900	Wallace, John Findlay
1874–75	Adams, Julius Walker	1901	Croes, John James Robertson
1876–77	Greene, George Sears	1902	Moore, Robert
1878	Chesbrough, Ellis Sylvester	1903	Noble, Alfred
1879	Roberts, William Milnor	1904	Hermany, Charles
1880	Fink, Albert	1905	Schneider, Charles Conrad
1881	Francis, James Bicheno	1906	Stearns, Frederic Pike
1882	Welch, Ashbel	1907	Benzenberg, George Henry
1883	Paine, Charles	1908	MacDonald. Charles
1884	Whittemore, Don Juan	1909	Bates, Onward
1885	Graff, Frederic	1910	Bensel, John Anderson
1886	Flad, Henry	1911	Endicott, Mordecai Thomas
1887	Worthen, William Ezra	1912	Ockerson, John Augustus
1888	Keefer, Thomas Coltrin	1913	Swain, George Fillmore
1889	Becker, Max Joseph	1914	McDonald, Hunter
1890	Shinn, William Powell	1915	Marx, Charles David
1891	Chanute, Octave	1916	Corthell, Elmer Lawrence
1892	Cohen, Mendes	1916	Herschel, Clemens
1893	Metcalf, William	1917	Pegram, George Herndon
1894	Craighill, William Price	1918	Talbot, Arthur Newell
1895	Morison, George Shattuck	1919	Curtis, Fayette Samuel

1920 Davis, Arthur Powell
1921 Webster, George Smedley
1922 Freeman, John Ripley
1923 Loweth, Charles Frederick
1924 Grunsky, Carl Edward
1925 Ridgway, Robert
1926 Davison, George Stewart
1927 Stevens, John Frank
1928 Bush, Lincoln
1929 Martson, Anson
1930 Coleman, John Francis
1931 Stuart, Francis Lee
1932 Crocker, Herbert Samuel
1933 Hammond, Alonzo John
1934 Eddy, Harrison Prescott
1935 Tuttle, Arthur Smith
1936 Mead, Daniel Webster
1937 Hill, Louis Clarence
1938 Riggs, Henry Earle
1939 Sawyer, Donald Hubbard
1940 Hogan, John Philip
1941 Fowler, Frederick Hall
1942 Black, Ernest Bateman
1943 Whitman, Ezra Bailey
1944 Pirnie, Malcolm
1945 Stevens, John Cyprian
1946 Horner, Wesley Winans
1947 Hastings, Edgar Morton
1948 Dougherty, Richard Erwin
1949 Thomas, Franklin
1950 Howard, Ernest Emmanuel
1951 Hathaway, Gail Abner
1952 Proctor, Carlton Springer
1953 Huber, Walter Leroy
1954 Terrell, Daniel Voiers
1955 Glidden, William Roy
1956 Needles, Enoch Ray
1957 Lockwood, Mason Graves
1958 Howson, Louis Richard
1959 Friel, Francis de Sales
1960 Marston, Frank Alwyn
1961 Holcomb, Glenn Willis
1962 Earnest, George Brooks
1963 Friedman, Edmund
1964 Bowman, Waldo Gleason
1965 Chadwick, Wallace Lacy
1966 Hedley, William Joseph
1967 Andrews, Earle Topley
1968 Tatlow, Richard Henry, III
1969 Newnam, Frank Hastings, Jr.
1970 Niles, Thomas McMaster
1971 Baxter, Samuel Serson
1972 Bray, Oscar Simon
1973 Rinne, John Elmer
1974 Yoder, Charles William
1975 Sangster, William McCoy
1976 Fox, Arthur Joseph, Jr.
1977 Walker, Leland Jasper
1978 Gibbs, William Read
1979 Blessey, Walter Emanuel
1980 Ward, Joseph Simeon
1981 Mendenhall, Irvan Frank
1982 Sims, James Redding
1983 Wiedeman, John H.
1984 Stearns, S. Russell
1985 Karn, Richard W.
1986 Bay, Robert Dewey
1987 Bargewood, Daniel Bythewood, Jr.
1988 Grant, Albert Abraham
1989 Carroll, William J.
1990 Focht, John A., Jr.
1991 Sawyer, James E. "Tom"
1992 Pennoni, Celestino R. "Chuck"
1993 McCarty, James E.
1994 Poirot, James W.
1995 Thornton, Stafford E.
1996 Parthum, Charles A.
1997 Groff, Edward O.
1998 Graef, Luther W.
1999 Turner, Daniel S.
2000 Hampton, Delon
2001 Bein, Robert W.
2002 Schwartz, H. Gerard, Jr.

Index